CONTEMPORARY MATHEMATICS

Plane Ellipticity and Related Problems

AMERICAN MATHEMATICAL SOCIETY

VOLUME 11

CONTEMPORARY MATHEMATICS

Titles in this Series

VOLUME 1 **Markov random fields and their applications**
Ross Kindermann and J. Laurie Snell

VOLUME 2 **Proceedings of the conference on integration, topology, and geometry in linear spaces**
William H. Graves, Editor

VOLUME 3 **The closed graph and P-closed graph properties in general topology**
T. R. Hamlett and L. L. Herrington

VOLUME 4 **Problems of elastic stability and vibrations**
Vadim Komkov, Editor

VOLUME 5 **Rational constructions of modules for simple Lie algebras**
George B. Seligman

VOLUME 6 **Umbral calculus and Hopf algebras**
Robert Morris, Editor

VOLUME 7 **Complex contour integral representation of cardinal spline functions**
Walter Schempp

VOLUME 8 **Ordered fields and real algebraic geometry**
D. W. Dubois and T. Recio, Editors

VOLUME 9 **Papers in algebra, analysis and statistics**
R. Lidl, Editor

VOLUME 10 **Operator algebras and K-theory**
Ronald G. Douglas and Claude Schochet, Editors

VOLUME 11 **Plane ellipticity and related problems**
Robert P. Gilbert, Editor

CONTEMPORARY MATHEMATICS

Volume 11

Plane Ellipticity and Related Problems

AMERICAN MATHEMATICAL SOCIETY

Providence • Rhode Island

PROCEEDINGS OF THE SPECIAL SESSION ON
ELLIPTIC SYSTEMS IN THE PLANE
87TH ANNUAL MEETING OF THE AMERICAN MATHEMATICAL SOCIETY

HELD IN SAN FRANCISCO, CALIFORNIA

JANUARY 7-11, 1981

EDITED BY
ROBERT P. GILBERT

Library of Congress Cataloging in Publication Data

Special Session on Elliptic Systems in the Plane (1981: San Francisco, Calif.)
Plane ellipticity and related problems.
(Contemporary mathematics, ISSN 0271-4132; v. 11)
"Proceedings of the Special Session on Elliptic Systems in the Plane, 87th Annual Meeting of the American Mathematical Society, held in San Francisco, California, January 7–11, 1981" —Verso t.p.
Bibliography: p.
1. Differential equations, Elliptic—Congresses. 2. Functions of complex variables—Congresses. I. Gilbert, Robert P., 1932– . II. American Mathematical Society. Meeting (87th: 1981: San Francisco, Calif.) III. Title. IV. Series: Contemporary mathematics (American Mathematical Society); v. 11.
QA377.S67 1981 515.3'53 82-11562
ISBN 0-8218-5012-1

1980 Mathematics Subject Classification. Primary 35J55, 35A92, 30A97.

CONTENTS

INTRODUCTION

In this collection of papers concepts associated with plane ellipticity is extended in several ways. For example, the investigations of Begehr and Gilbert, Begehr and Hsiao, Hile, and Snyder treat systems of elliptic partial differential equations in the plane which resemble in some sense the Cauchy-Riemann equations. Their point of view is to seek general representation formulas and to use these in some cases to solve boundary value problems. Continuing with the theme of generalizing the Cauchy-Riemann equations Buchanan treats the Bers-Vekua type systems in two complex variables, while Delanghe and Sommen, Brackx and Pincket, and Lounesto investigate hypercomplex function theory in $\mathbf{R}^n$, that is the class of monogenic functions having values in a Clifford algebra.

The remaining talks comprising this special meeting cannot be categorized as falling into a general group, but rather explore isolated, albeit important topics associated with ellipticity. In the collection we have the paper by Treves which answers a fundamental question posed by Lee Rubel, namely does there exist a real homogeneous, linear partial differential equation in a domain of $\mathbf{R}^2$ ($\mathbf{R}^n$) having only the trivial solution as a solution. Another work in this general group was a paper by Protter, who showed that Payne's method of obtaining gradient bounds by means of the maximum principle may be extended to a fairly general class of semilinear elliptic equations.

The manuscripts by Bruch and Sloss, and Hummel are involved with the method of variational inequalities for problems of planar fluid flow. Hummel investigates these flows by means of the hodograph transformation, whereas Bruch and Sloss treat a free-boundary problem by first introducing the Baiocchi transformation.

Finally we have the papers by McCoy and Aziz, Door and Kellogg. McCoy uses integral operators to generate a family of basis functions in order to obtain an analogue of the Favard-Achieser-Krein approximation theorem. The paper by Aziz et. al. develops the finite element method for an exterior boundary value problem associated with a system of elliptic partial differential equations which occur in the scattering of electromagnetic waves.

Robert P. Gilbert

Contemporary Mathematics
Volume 11, 1982

THE FINITE ELEMENT METHOD AND NON-LOCAL BOUNDARY CONDITIONS FOR SCATTERING PROBLEMS

by

A. K. Aziz
Department of Mathematics, University of Maryland, Baltimore County, and Institute for Physical Science and Technology, University of Maryland, College Park.

and

M. R. Dorr
Naval Surface Weapons Center, Silver Spring, MD.

and

R. B. Kellogg
Institute for Physical Science and Technology,
University of Maryland, College Park

1. Introduction

In this paper we describe a unified approach to the numerical approximation of the solutions to a class of scattering problems for the reduced wave and Maxwell's equations.

The numerical calculation of scattering by an arbitrarily shaped perfect conductor or dielectric body has received much attention in recent years. The problem is important in many areas including the design of waveguides and missile fuzes, the assessment of damage by an electromagnetic pulse, and the study of the biological effects of microwave radiation. Mathematically, the problem has the form of an exterior boundary value problem for a system of elliptic partial differential equations. The problem has particular difficulties arising from the large number of unknowns, the unbounded domain, and the fact that the solution is oscillatory for large values of frequency.

The principal techniques for the numerical solution of scattering problems are the method of integral equations and the finite difference or finite element method. For more details we refer to [1], [2], [8], [10], [11], [20], [21], [22] and the references contained therein.

Here we present a new finite element procedure where the non-local nature of the boundary conditions plays an essential role. Our approximation method possesses the following features.

(i) The subspaces are a combination of finite element subspaces inside the absorbing domain and spherical harmonics in the exterior domain. In addition to introducing a novel way of coupling the spherical harmonics with the field inside the body, the method has the feature that no finite elements are required outside the absorbing body. Moreover, in the case of a perfect conductor, only the spherical harmonics in the exterior domain are required. (ii) The stiffness matrix and each of its principal minors is non-singular. Thus, a sparse matrix solver may be used to save computer storage. In [1] this method is analyzed for the reduced wave equation when Ω is an absorbing domain. The case when Ω is a perfectly conducting domain is treated in [3]. The extension of the method to the reduced Maxwell's equations is given in [2].

0271-4132/82/0000-0139/$04.00

In section 2 we formulate our problem for the reduced wave equation in the case Ω is an absorbing domain; section 2.1 contains a series of results (whose proofs appear in [1]) which lead to our fundamental Theorem 2.1. In section 2.2 we describe the finite element method for our problem and give an error analysis and a brief discussion of the finite dimensional subspaces involved. In section 3 we consider the case when Ω is a perfectly conducting domain. Section 4 deals with the extension of the method to the reduced Maxwell's equations. This paper essentially gives a survey of the results contained in [1], [2], [3].

2. Formulation of the Problem for an Absorbing Domain

Let Ω be a bounded domain in $\mathbb{R}^3$ with smooth boundary Γ, and let $\Omega_o = \mathbb{R}^3 \setminus \bar{\Omega}$ be the exterior domain. Denote by n the unit normal to Γ pointing out of Ω. Let $k(x)$ be a bounded, complex-valued function on $\mathbb{R}^3$ such that

$$(2.1)\qquad \begin{cases} \operatorname{Im} k(x) \geq Q > 0, & x \in \Omega \\ k(x) = k_o = \text{real constant}, & x \in \Omega_o \end{cases}.$$

Finally, let $f(x)$ be a function defined on $\mathbb{R}^3$ with $f(x) = 0$ for $x \in \Omega_o$, and let u_o be a function satisfying

$$(2.2)\qquad \Delta u_o(x) + k_o^2 u_o(x) = 0 \qquad x \in \mathbb{R}^3 .$$

Our goal is to solve the following problem:

Problem 1. Find $u \in H^1_{loc}(\mathbb{R}^3)$ satisfying:

$$(2.3)\qquad \Delta u + k^2 u = f \quad \text{in } \mathbb{R}^3 \setminus \Gamma$$

$$(2.4)\qquad \begin{cases} u - u_o = O(r^{-1}), \ r = |x| \to \infty \\ \frac{\partial}{\partial r}(u-u_o) - ik_o(u-u_o) = o(r^{-1}), \ r = |x| \to \infty \end{cases}.$$

We deal with Problem 1 by introducing an equivalent variational problem on the bounded domain Ω.

2.1 Variational formulation

We first introduce the auxiliary problem $II_{\pm}$: let $s \in \mathbb{R}^1$, $g \in H^s(\Gamma)$ be given. Find $v_{\pm} \in H^{s+\frac{1}{2}}_{loc}(\Omega_o)$ such that

$$(2.1.1)\qquad \Delta v_{\pm} + k_o^2 v_{\pm} = 0 \quad \text{in } \Omega_o$$

$$(2.1.2)\qquad \gamma v_{\pm} = g \quad \text{on } \Gamma$$

$$(2.1.3)\qquad \begin{cases} v_{\pm} = O(r^{-1}), & |x| = r \to \infty \\ \frac{\partial v_{\pm}}{\partial r} \mp ik_o v_{\pm} = o(r^{-1}), & |x| = r \to \infty \end{cases},$$

where γ denotes the usual trace operator restricting functions on Ω_o to Γ. It is well known [16] that the problem (2.1.1)-(2.1.3) has a unique solution and that the normal derivative $\gamma \frac{\partial v_{\pm}}{\partial n} \in H^{s-1}(\Gamma)$. We define $K_{\pm}$ to be the

the mapping given by:

$$K_{\pm}g \equiv \gamma \frac{\partial v_{\pm}}{\partial n} \quad ,$$

where $v_{\pm}$ is the unique solution of problem $II_{\pm}$. Thus for each $s \in \mathbb{R}^1$, $K_{\pm}$ is a bounded map from $H^{s}(\Gamma)$ into $H^{s-1}(\Gamma)$. We also have that $K_{\pm}$ is a pseudo-differential operator on $H^{s}(\Gamma)$ of order 1 (see [1]).

We now reformulate problem (2.3)-(2.4) by introducing a bilinear form on $H^1(\Omega) \times H^1(\Omega)$ as follows. Let $(\cdot,\cdot)$ (resp. $<\cdot,\cdot>$) denote the usual inner product in $L_2(\Omega)$ (resp. $L_2(\Gamma)$). We also let $<\cdot,\cdot>$ denote the duality pairing between $H^{s}(\Gamma)$ and $H^{-s}(\Gamma)$. We define a bilinear form B by

$$\text{(2.1.4)} \quad B(u,v) = -(\nabla u\cdot\nabla v) + (k^2u,v) + <K_{\pm}\gamma u,v>, \ (u,v) \in H^1(\Omega) \times H^1(\Omega) \quad .$$

Using the bilinear form (2.1.4), we introduce Problem III: find $u \in H^1(\Omega)$ such that

$$\text{(2.1.5)} \qquad B(u,v) = B(u_o,v) + (f-\Delta u_o-k^2u_o,v), \qquad v \in H^1(\Omega) \quad .$$

It is easily seen that Problem III yields a reformulation of Problem I. In fact, suppose u solves Problem I and $u \in H^2(\Omega)$. Letting $z = u-u_o$ and using Green's formula we have

$$(f-\Delta u_o-k^2u_o,v) = (\Delta z+k^2z,v) = -(\nabla z\cdot\nabla v)+(k^2z,v)+<z_n,v>$$

Since, in Ω_o, z solves Problem $II_{\pm}$ with $g = \gamma z$, it follows that $z_n = K_{\pm}z$, and we have (2.1.5).

Let

$$H^2_{\pm}(\Omega) = \{u \in H^2(\Omega): u_n=K_{\pm}\gamma u\} \quad .$$

We observe that $H^2_{\pm}(\Omega)$ is a closed subspace of $H^2(\Omega)$. We define the mappings $A_{\pm}: H^2_{\pm}(\Omega) \to H^0(\Omega)$ by

$$A_+u = \Delta u+k^2u, \quad A_-u = \Delta u+\bar{k}^2u \quad .$$

The operators $A_{\pm}$, regarded as unbounded operators on $L^2(\Omega)$ with domains $H^2_{\pm}(\Omega)$, have the following properties (see [1]):

(i) $(A_{\pm}u,v) = B(u,v), \quad u \in H^2_{\pm}(\Omega), \ v \in H^1(\Omega)$,

(ii) $(A_{+}u,v) = (u,A_-v), \quad u \in H^2_{+}(\Omega), \quad v \in H^2_{-}(\Omega)$,

(iii) $\alpha||u||_{H^o(\Omega)} \leq ||A_{\pm}u||_{H^o(\Omega)}, \qquad u \in H^2_{\pm}(\Omega)$

(iv) $A_{\pm}$ are closed, densely defined, invertible operators on $L^2(\Omega)$ and $(A_{\pm})^* = A_{\mp}$.

The equation $u_n = K_{\pm}\gamma u$, which defines the space $H^2_{\pm}(\Omega)$, may be considered as a boundary condition of non-local type. Grisvard [10] has given a general result on the interpolation between the subspaces of Sobolev spaces defined by boundary conditions of local type, e.g., differential operators. A result of this type holds for our spaces $H^2_{\pm}(\Omega)$. More precisely we have (see [1]):

(2.1.6) $$[H_{\pm}^{2}(\Omega),L_{2}(\Omega)]_{\frac{1}{2}} = H^{1}(\Omega) .$$

Now we are in a position to state and prove the main result of this section:

Theorem 2.1. There exists a constant $C > 0$ such that for each $u \in H^{1}(\Omega)$ there is a $v \in H^{1}(\Omega)$ satisfying the inequality

$$|B(u,v)| \geq C||u||_{H^{1}(\Omega)}||v||_{H^{1}(\Omega)} .$$

Proof. Regarding $A_{\pm}$ as a bounded, invertible map of $H_{\pm}^{2}(\Omega) \to L_{2}(\Omega)$, we extend A_{+} to a map of $L_{2}(\Omega) \to H_{-}^{2}(\Omega)$ as follows: if $f \in L_{2}(\Omega)$ consider $A_{+}f$ as a linear functional on $H^{2}(\Omega)$ given by

$$(A_{+}f)(v) = (f,A_{-}v) .$$

In particular, if $f \in H_{+}^{2}(\Omega)$, we have $(A_{+}f)v = (A_{+}f,v)$, so this definition agrees with the previous definition. The extended map is 1-1 and invertible. By interpolation we have that

$$A_{+}: [H_{+}^{2}(\Omega),L_{2}(\Omega)]_{\frac{1}{2}} \to [L^{2}(\Omega),H^{2}(\Omega)]_{\frac{1}{2}}$$

is bounded, 1-1, and invertible.

From (2.1.6) and duality properties of interpolation it follows that

$$A_{+}: H^{1}(\Omega) \to H^{1}(\Omega)'$$

is a bounded, 1-1, invertible map. From the property (i) above we find that

$$(A_{+}u,v) = B(u,v), \quad u,v \in H^{1}(\Omega) .$$

Let $v \in H^{1}(\Omega)$ be given. Choose $\xi \in H^{1}(\Omega)'$ such that

$$\xi(v) = ||v||_{H^{1}(\Omega)}, \quad ||\xi||_{H^{1}(\Omega)'} = 1 .$$

Let $u = A_{+}^{-1}\xi \in H^{1}(\Omega)$. Then

$$B(u,v) = ||v||_{H^{1}(\Omega)} \geq C||v||_{H^{1}(\Omega)}||u||_{H^{1}(\Omega)'}$$

with $C = ||A_{+}^{-1}||$, where the norm refers to the map

$$A_{+}^{-1}: H^{1}(\Omega)' \to H^{1}(\Omega) .$$

We shall need in the sequel a Garding type inequality for our bilinear form. Namely, for $u \in H^{1}(\Omega)$

(2.1.7) $$|B(u,u)| \geq C_{1}||u||_{H^{1}(\Omega)}^{2} - C_{2}||u||_{H^{\theta}(\Omega)}^{2}, \quad \frac{1}{2} < \theta < 1 ,$$

with $C_{i} > 0$, $i = 1,2$. (For a proof see [1].)

2.2 Discrete Problem

Let $S \subset H^{1}(\Omega)$ be finite dimensional. In analogy with (2.1.5) we seek $\tilde{u} \in S$ such that

(2.2.1) $$B(\tilde{u},v) = B(u_{o},v) + (f-\Delta u_{o}-k^{2}u_{o},v), \quad v \in S .$$

We shall refer to $\tilde{u}$ as the approximate solution of Problem I, using the subspace S. To see that $\tilde{u}$ is well defined, we note that the equations (2.2.1), when expressed in terms of a basis for S, yield a finite system of linear equations which is nonsingular. In fact, if $z \in S$ and $B(z,v) = 0$ for all $v \in S$, then $z = 0$, since by choosing $v = z$ and applying (iii), we have

$$\alpha||z||^2 \leq |B(z,z)| = 0 \quad .$$

To give an analysis of the discretization error $u-\tilde{u}$, we shall show that B satisfies a discrete form of the inf-sup condition. To this end we first state a weak form of the inf-sup condition which may be also of use in other problems (see Schatz [17] for a similar result).

Lemma 2.2. Let H_i, $i = 0,1,2$, be three Hilbert spaces. Suppose $H_0 \supset H_1$ with compact injection. Let B be a bounded bilinear form on $H_1 \times H_2$ which satisfies: if $u \in H_1$ and $B(u,v) = 0$, $v \in H_2$, then $u = 0$. For each $N, N-1, 2, \ldots$, let $M_{i,N} \subset H_i$, $i = 1,2$, be two finite dimensional subspaces of equal dimension such that $M_{2,N} \subset M_{2,N+1}$ and such that $\bigcup_N M_{2,N}$ is dense in H_2. Suppose B satisfies the "weak inf-sup condition": there are constants $C_i > 0$, $i = 1,2$ such that for $u \in M_{1,N}$, there is a $v \in M_{2,N}$ satisfying

$$|B(u,v)| \geq [C_1||u||_{H_1} - C_2||u||_{H_0}]||v||_{H_2} . \tag{2.2.2}$$

Then there is an integer $N_0 > 0$ and a constant $C_3 > 0$ such that for $N \geq N_0$, if $u \in M_{1,N}$, there is a $v \in M_{2,N}$ satisfying

$$|B(u,v)| \geq C_3||u||_{H_1}||v||_{H_2} \quad . \tag{2.2.3}$$

For a proof of lemma 2.2 we refer to [1].

As an immediate consequence of lemma 2.2 and the uniqueness of the solution of the system (2.2.1), we have (see [1])

Theorem 2.2. Let $S^j \subset H^1(\Omega)$, $j = 1,2,\ldots$, be an increasing family of finite dimensional subspaces of $H^1(\Omega)$ such that $\bigcup S^j$ is dense in $H^1(\Omega)$. Let $u_j \in S^j$ be the approximate solution of Problem I using the subspace S^j. Then there is a constant $C > 0$, independent of j but depending on the family $\{S^j\}$, such that if u is a solution of Problem I,

$$||u-u_j||_{H^1(\Omega)} \leq C \inf\{||u-z||_{H^1(\Omega)} : z \in S^j\} \quad . \tag{2.2.4}$$

Equation (2.2.4) indicates that our approximate method gives, in a quasi-optimal sense, as good an approximation to the solution as can be expected from the subspace that is being used.

To find the approximate solution $\tilde{u}$ using a subspace S, we choose a basis $\{z_i\}$, $1 \leq i \leq m$, of S. Letting

$$A = [a_{ij}], \quad a_{ij} = B(z_i,z_j), \quad \underline{F} = [f_i] \quad ,$$

$$f_i = B(u_o,z_i) + (f-\Delta u_o-k^2u_o,z_i), \quad 1 \leq i \leq m,$$

and writing $u(x) = \sum u_i z_i(x)$, $\underline{U} = [u_i]$, (2.2.1) may be written as the matrix system $A\underline{u} = \underline{F}$. As we saw earlier, this equation always has a solution. It is shown in [1] that the matrix A can be factored without pivoting, so the unknowns can be arranged to minimize the storage requirements of the matrix.

Our approximate method has a potential difficulty in that the operator

K_+, and thus the bilinear form B, is difficult to evaluate. However, this difficulty can be overcome by an appropriate choice of subspaces, which we now describe briefly. Let $S_N \subset H^1_{loc}(\Omega_o)$, $N = 1,2,\ldots$, be increasing sequence of finite dimensional subspaces of functions which satisfy

(2.2.5) $$\Delta v + k_o^2 v = 0 \quad \text{in } \Omega_o$$

(2.2.6) $$\begin{cases} v(x) = O(r^{-1}), & r = |x| \to \infty \\ \dfrac{\partial v(x)}{\partial r} - ik_o v(x) = o(r^{-1}), & r \to \infty \end{cases}$$

A specific choice of S_N arises, for example, from the separation of variables in spherical coordinates applied to (2.2.5). Suppose $0 \in \Omega$. Let $Y_{m,n}(\theta,\phi)$ be a surface harmonic, and let $h^1_n(\rho)$ be a spherical Bessel function [19, chapter 10]. Then $v(r,\theta,\phi) = h^1_n(k_o r)Y_{m,n}(\theta,\phi)$ is a particular solution of (2.2.5), (2.2.6). S_N may be taken to be the collection of all such solutions v with $n = 1,\ldots,N$, $m = 0, \pm 1,\ldots,\pm n$. For $v \in S_N$, $K\gamma v = \gamma\, \partial v/\partial n$ can easily be calculated.

We also need a collection of functions of finite element type. Let there be given a decomposition of $\mathbb{R}^3$ into simplices of maximum size h. Let V^h be the set of continuous piecewise linear functions on this triangulation. Let $V^h_o = V_h \cap H^1_o(\Omega)$. The subspaces of function used in our variational formulation is obtained by combining the spaces V^h_o and S_N. For more details on this, see [1].

3. Formulation of the Problem for a Perfectly Conducting Domain

As in the previous section let $\Omega \subset \mathbb{R}^3$ be a bounded domain with smooth boundary Γ and let $\Omega_o = \mathbb{R}^3 \setminus \bar{\Omega}$ be the exterior domain. For a given $f \in H^{\frac{1}{2}}(\Omega)$ and k a real, positive number, we consider the problem: find $u \in H^1_{loc}(\Omega_o)$ such that

(3.1) $$\Delta u + k^2 u = 0 \quad \text{in } \Omega_o$$

(3.2) $$\gamma u = f \quad \text{on } \Gamma ,$$

(3.3) $$\begin{cases} u = O(r^{-1}) & \text{as } |x| = r \to \infty \\ \dfrac{\partial u}{\partial r} - iku = o(r^{-1}) & \text{as } |x| = r \to \infty , \end{cases}$$

where γ again denotes the usual trace operator on Γ.

It is known [3], [9] that for any $f \in H^{\frac{1}{2}}(\Gamma)$ there exists a unique solution of (3.1), (3.2), (3.3) in $H^1_{loc}(\Omega_o)$ and that the normal derivative $\frac{\partial u}{\partial n} \in H^{-\frac{1}{2}}(\Gamma)$. Moreover, if we let $K\colon H^{\frac{1}{2}}(\Gamma) \to H^{-\frac{1}{2}}(\Gamma)$ be defined by: $Kf = \gamma \frac{\partial u}{\partial n}$ then K is an isomorphism of $H^{\frac{1}{2}}(\Gamma)$ onto $H^{-\frac{1}{2}}(\Gamma)$.

3.1. A variational formulation

Let f and g be any two elements of $H^{\frac{1}{2}}(\Gamma)$. We define a bilinear form on $H^{\frac{1}{2}}(\Gamma) \times H^{\frac{1}{2}}(\Gamma)$ by

$$B(f,g) = \langle f, Kg \rangle ,$$

where $\langle \cdot,\cdot \rangle$ denotes the sesquilinear form defining the $H^{\frac{1}{2}}(\Gamma) - H^{-\frac{1}{2}}(\Gamma)$ duality pairing, and K is the operator defined above. From the fact that K is an isomorphism of $H^{\frac{1}{2}}(\Gamma)$ onto $H^{-\frac{1}{2}}(\Gamma)$, it follows that B is both well defined and bounded on $H^{\frac{1}{2}}(\Gamma) \times H^{\frac{1}{2}}(\Gamma)$. Furthermore B satisfies a Garding type inequality (see [3]) which we state as

Theorem 3.1.1. There exist constants C_1 and C_2 such that for $f \in H^{\frac{1}{2}}(\Gamma)$

$$|B(f,f)| \geq C_1 ||f||^2_{H^{\frac{1}{2}}(\Gamma)} - C_2 ||f||^2_{H^{\theta_o}(\Gamma)} \tag{3.1.1}$$

for some θ_o, $0 < \theta_o < \frac{1}{2}$.

We now reformulate problem (3.1), (3.2), (3.3) as a variational problem on Γ involving the bilinear form B introduced above. To this end, let u_f denote the solution of (3.1), (3.2), (3.3) with $f \in H^{\frac{1}{2}}(\Gamma)$. Define

$$S = \{u_f : f \in H^{\frac{1}{2}}(\Gamma)\} \subset H^1_{loc}(\Omega_o) .$$

We consider the problem: given $f \in H^{\frac{1}{2}}(\ )$, find $u \in S$ such that

$$B(\gamma u, \gamma v) = \langle f, \gamma \frac{\partial v}{\partial n} \rangle \quad \forall v \in S . \tag{3.1.2}$$

Since $\gamma \frac{\partial v}{\partial n} = K\gamma v$ for $v \in S$, and since K is an isomorphism of $H^{\frac{1}{2}}(\Gamma)$ onto $H^{-\frac{1}{2}}(\Gamma)$, it follows that (3.1.2) has the unique solution $u = u_f \in S$. Therefore (3.1.2) gives an equivalent formulation of (3.1), (3.2), (3.3). The advantage of this formulation will become clear in the next section where we discretize (3.1.2) and obtain a quasi-optimal error estimate for the resulting approximate solution.

3.2 The discretized problem and error estimate

In this section we develop a numerical procedure based on the bilinear form B for the approximation of the solution to problem (3.1), (3.2), (3.3). To this end, we discretize the variational problem (3.1.2). Let S_N be a finite dimensional subspace of S. Then for a given $f \in H^{\frac{1}{2}}(\Gamma)$ (3.1.2) is replaced by the following discrete analogue: find $u_N \in S_N$ such that:

$$B(\gamma u_N, \gamma v_N) = \langle f, \gamma \frac{\partial v_N}{\partial n} \rangle \quad \forall v_N \in S_N . \tag{3.2.1}$$

By expressing u_N and v_N in terms of a basis for S_N, (3.2.1) gives a finite system of linear equations whose solution yields the coefficients of u_N with respect to the chosen basis.

Clearly, in order to implement (3.2.1), one needs to know explicitly the space S_N, and it must be possible to compute $K\gamma v$ for any $v \in S_N$. This can be achieved by a suitable choice of S_N. As in section 2.2 we consider an increasing sequence $\{S_N\}$ of finite dimensional subspaces of $H^1_{loc}(\Omega_o)$ consisting of functions v which satisfy:

$$\Delta v + k^2 v = 0 \tag{3.2.2}$$

$$\begin{cases} v(x) = O(r^{-1}) & r = |x| \to \infty \\ \dfrac{\partial v(x)}{\partial r} - ikv(x) = o(r^{-1}), & r = |x| \to \infty \end{cases} \tag{3.2.3}$$

A specific choice of S_N was given in section 2.2. For more on this we refer to [3].

Next we give an error analysis for our approximate method. In this connection we have the following theorem.

Theorem 3.2.1. Let S_N be an increasing family of finite dimensional subspaces of $H^1_{loc}(\Omega_o)$ consisting of solutions of (3.2.2), (3.2.3). Let $S = \bigcup_N S_N$, and suppose γS is dense in $H^{\frac{1}{2}}(\Gamma)$. Then for N sufficiently large there exists a unique solution $u_N \in S_N$ of (3.2.1). Moreover if u is a solution of (3.1), (3.2), (3.3) then

$$(3.2.4) \qquad ||\gamma u-\gamma u_N||_{H^{\frac{1}{2}}(\Gamma)} \leq C \inf_{v_N \in S_N} ||\gamma u-\gamma v_N||_{H^{\frac{1}{2}}(\Gamma)} ,$$

where the constant $C > 0$ is independent of N but depends on the family $\{S_N\}$.

The proof of theorem 3.2.1 is similar to the proof of theorem 2.2. In fact, from theorem 3.1.1, it follows that the hypotheses of lemma 2.2 are satisfied with $H_1 = H_2 = H^{\frac{1}{2}}(\Gamma)$, $H_0 = H^{\theta_o}(\Gamma)$, $0 < \theta_o < \frac{1}{2}$, $M_{1,N} = M_{2,N} = \{\gamma u: u \in S_N\}$. Using the analogue of (2.2.3) we have

$$|B(f,g)| \geq C||f||_{H^{\frac{1}{2}}(\Gamma)} ||g||_{H^{\frac{1}{2}}(\Gamma)} .$$

Thus the hypotheses of [4, theorem 6.2.1] hold. Hence it follows that there exists a unique solution $u_N \in S_N$ of (3.2.1) and moreover there is a constant $C > 0$ such that

$$||\gamma u-\gamma u_N||_{H^{\frac{1}{2}}(\Gamma)} \leq C \inf_{v_N \in S_N} ||\gamma u-\gamma v_N||_{H^{\frac{1}{2}}(\Gamma)} .$$

Remark 3.1. If $G \subset \mathbb{R}^3$ is compact, (3.2.4) implies

$$||u-u_N||_{H^1(\Omega_o \cap G)} \leq C \inf_{v_N \in S_N} ||\gamma u-\gamma v_N||_{H^{\frac{1}{2}}(\Gamma)} ,$$

where the constant $C > 0$ depends on the set G and the family $\{S_N\}$, but it is independent of N. If $G \supset \Gamma$, we have the stronger inequality

$$||u-u_N||_{H^1(\Omega_o \cap G)} \leq C \inf_{v_N \in S_N} ||u-v_N||_{H^1(\Omega_o \cap G)} .$$

Remark 3.2. In [3] the density property assumed in theorem 3.2.1 is shown to be satisfied by the subspaces S_N generated by the spherical harmonics described in section 2.

4. Reduced Maxwell's Equation

In sections 2 and 3 we developed an approximation method for the solution of scattering problem for the reduced wave equation. In the present section we briefly describe the application of the method given in section 3 to the reduced Maxwell's equations in a perfectly conducting domain.

As in section 3, let $\Omega \subset \mathbb{R}^3$ be a bounded perfectly conducting domain with boundary Γ. We allow the possibility that Ω is a disconnected domain, in which case Γ is a collection of disjoint closed surfaces. Let $\Omega_o = \mathbb{R}^3 \setminus \bar{\Omega}$ denote the complement of Ω in $\mathbb{R}^3$. Let $\underline{E}^o$ and $\underline{H}^o$ be two vector fields which satisfy in $\mathbb{R}^3$ the reduced Maxwell's equations

$$\underline{\nabla} \times \underline{E}^o = \alpha \underline{H}^o, \qquad \underline{\nabla} \times \underline{H}^o = \beta \underline{E}^o ,$$

where $\alpha = i\alpha_o$, $\beta = -i\beta_o$, with $\alpha_o > 0$, $\beta_o > 0$. Let $\underline{E}$, $\underline{H}$ be the wave resulting from the scattering of the incident wave $\underline{E}^o$, $\underline{H}^o$ by the perfectly conducting body Ω. Let $\underline{E}^1 = \underline{E} - \underline{E}^o$, $\underline{H}^1 = \underline{H} - \underline{H}^o$ be the scattered wave. Then $\underline{E}(x)$ and $\underline{H}(x)$ are defined for $x \in \Omega_o$, and satisfy the following set of equations:

(4.1a) $\quad \underline{\nabla} \times \underline{E} = \alpha \underline{H} \qquad x \in \Omega_o$

(4.1b) $\quad \underline{\nabla} \times \underline{H} = \beta \underline{E} \qquad x \in \Omega_o$

(4.2) $\quad \underline{E} \times \underline{n} = 0 \qquad x \in \Gamma$

(4.3a) $\quad \underline{E}^1, \underline{H}^1 = O(r^{-1}) \qquad r = |x| \to \infty$

(4.3b) $\quad \underline{e}_r \times \underline{E}^1 - \sqrt{\frac{\alpha_o}{\beta_o}}\,\underline{H}^1 = o(r^{-1}) \qquad r = |x| \to \infty$

(4.3c) $\quad \underline{e}_r \times \underline{H}^1 + \sqrt{\frac{\beta_o}{\alpha_o}}\,\underline{E}^1 = o(r^{-1}) \qquad r = |x| \to \infty$

Here $\underline{n}$ denotes the outward pointing unit normal to Γ and $\underline{e}_r$ denotes the unit vector in the radial direction.

The system (4.1a,b) comprises a system of six partial differential equations in the six unknowns consisting of the components of $\underline{E}$ and $\underline{H}$. The equations (4.1), (4.2), and (4.3) form an exterior boundary value problem for this system; (4.2) is the boundary condition on Γ, and (4.3) are boundary conditions at infinity. This has been treated, e.g., in [16], where it is shown that, for reasonable surfaces Γ and incident waves $\underline{E}^o$, $\underline{H}^o$, the problem has a unique solution.

4.1 Variational formulation.

It is convenient to discuss the boundary value problem in terms of the scattered wave. For this let $\underline{f}$ be a tangential vector field on Γ, and consider the boundary value problem:

(4.1.1a) $\quad \underline{\nabla} \times \underline{E}^1 = \alpha \underline{H}^1 \qquad x \in \Omega_o$

(4.1.1b) $\quad \underline{\nabla} \times \underline{H}^1 = \beta \underline{E}^1 \qquad x \in \Omega_o$

(4.1.2) $\quad \underline{n} \times \underline{E}^1 = f \qquad x \in \Gamma$,

and (4.3). From [16], we know that the above problem has a unique solution. If $\underline{f} = -\underline{n} \times \underline{E}^o$, then the solution $\underline{E}^1$, $\underline{H}^1$ of (4.1.1), (4.1.2) and (4.3) gives the scattered wave for the original problem. Thus, it suffices to solve (4.1.1), (4.1.2) and (4.3).

Let $\underline{f}$ be a tangential vector field on Γ. We define another tangential vector field $K\underline{f}$ as follows. Let $\underline{E}^1$, $\underline{H}^1$ be the solution of (4.1.1), (4.1.2), (4.3), and let $K\underline{f} = \underline{n} \times \underline{H}^1$. The operator K maps tangential vector fields into tangential vector fields. We now define a bilinear form B on tangential vector fields. If $\underline{f}$, $\underline{g}$ are two tangential vector fields on Γ, we define

$$B(f,g) = \oint_\Gamma \underline{n} \cdot \underline{f} \times K g \, d\Gamma \quad .$$

The operator K and the bilinear form B play a central role in the formulation of our approximation scheme which will be described in the next section. We first state two important properties of our bilinear form whose proofs are given in [2].

i) $B(\underline{f},\underline{g}) = B(\underline{g},\underline{f})$, i.e., B is symmetric.

ii) If $B(\underline{f}, \overline{\underline{f}}) = 0$ then $f = 0$, i.e., B is definite. ($\overline{\underline{f}}$ denotes the complex conjugate of $\underline{f}$.)

4.2 Discrete problem

We describe our numerical method in terms of the scattered wave $\underline{E}^1$. Let $\underline{g}$ be any tangential vector field on Γ. Since $\underline{n} \times \underline{E}^1 = -\underline{n} \times \underline{E}^o$ on Γ, we have

$$\underline{n}\cdot\underline{E}^1\times K\underline{g} = \underline{n}\times\underline{E}^1\cdot K\underline{g} = -\underline{n}\times\underline{E}^o\cdot K\underline{g} = -\underline{n}\cdot\underline{E}^o\times K\underline{g} \quad .$$

Integrating this over Γ, and denoting by $\underline{f}$ the tangential projection of the restriction of $\underline{E}^1$ to Γ, we have

$$B(\underline{f},\underline{g}) = -\oint_\Gamma \underline{n}\cdot\underline{E}^o\times K\underline{g}\,d\Gamma \quad . \tag{4.2.1}$$

(4.2) is the basis for our numerical method. Let S be a finite dimensional collection of tangential vector fields on Γ. We define our approximate solution $\underline{f}\in S$ by

$$B(\underline{f},\underline{g}) = -\oint_\Gamma \underline{n}\cdot\underline{E}^o\times K\underline{g}\,d\Gamma \ , \qquad \underline{g}\in S \ . \tag{4.2.2}$$

The system (4.2.2) gives rise to a finite system of linear equations whose solution determines the vector field $\underline{f}\in S$.

This approximation scheme seems to have two defects. It is not clear how to obtain the approximate scattered field in Ω_o from $\underline{f}$, and it is not clear how to evaluate the operator K which appears in (4.2.2). These defects may be remedied by a judicious choice of subspace, which we now describe.

Let $x^*\in\Omega$ be given, and let (r,θ,ϕ) be a system of spherical coordinates with the origin at x^*. In [19, p.416], a family of vector fields,

$$\underline{m}_{\substack{e\\o}mn} \ , \quad \underline{n}_{\substack{e\\o}mn} \ , \quad m = 0,1,\ldots,n, \quad n = 1,2,\ldots, \tag{4.2.3}$$

are constructed with the following properties.

(a) The fields are regular for $x \neq x^*$, and hence are in Ω_o.

(b) The fields satisfy

$$\nabla \times \underline{m}_{\substack{e\\o}mn} = \sqrt{\alpha_o\beta_o}\,\underline{n}_{\substack{e\\o}mn} \ ,$$

$$\nabla \times \underline{n}_{\substack{e\\o}mn} = \sqrt{\alpha_o\beta_o}\,\underline{m}_{\substack{e\\o}mn} \ .$$

(c) The fields satisfy (4.3).

In the formulas in [19, p.416] we must take $z_n(\rho) = h^1{}_n(\rho)$ in order to satisfy (c), where $h^1{}_n(\rho)$ denotes, as usual, the spherical Bessel function of the third kind.

We fix an integer $N > 0$ and let C_N denote the collection of the vector fields (4.2.3) for $0 \le m \le n$, $1 \le n \le N$. Let S_N denote the collection of tangential vector fields on Γ which are of the form $\underline{n} \times (\gamma\underline{U})$ where $\underline{U}\in C_N$. There are $2N(N+2)$ linearly independent fields in S_N. If $\underline{g}\in S_N$, using (b) above, $K\underline{g}$ may be easily calculated. If $\underline{f}\in S_N$ is the solution of (4.2.2), then $\underline{f}$ comes from a vector field $\underline{U}\in C_N$; the field $\underline{U}$ is the desired approximate scattered field and may be easily evaluated at points of Ω_o. We have thus shown how to overcome the defects of using (4.2.2).

We now discuss the system of equations arising from the use of (4.2.2). We arrange the fields (4.2.3) of C_N in a definite order and denote them by $\underline{F}_\mu$, $1 \leq \mu \leq M = 2N(N+2)$. Let $\underline{f}_\mu = \underline{n} \times (\gamma \underline{F}_\mu)$. Thus, the $\underline{f}_\mu$, $1 \leq \mu \leq M$, are tangential vector fields on Γ which form a basis for S_N . Denoting the desired solution $\underline{f}$ of (4.2.2) as $\underline{f} = \sum c_\mu \underline{f}_\mu$, we obtain the linear system

$$(4.2.4) \qquad \sum_{\nu=1}^{M} c_\nu B(\underline{f}_\mu, \underline{f}_\nu) = - \oint_\Gamma n \cdot \underline{E}^o \times K \underline{f}_\mu d\Gamma, \qquad 1 \leq \mu \leq M .$$

If we set $a_{\mu\nu} = B(\underline{f}_\mu, \underline{f}_\nu)$ and let $A = [a_{\mu\nu}]$ denote the coefficient matrix, we see at once from the property i) of our bilinear form B that A is symmetric. Moreover property ii) implies that A and its principal minors are nonsingular. Now from the arguments of [18, Theorem 3.1] the existence of the Cholesky decomposition follows. Thus, we have:

Theorem 4.2.1. The complex matrix A is symmetric, nonsingular, has nonsingular principal minors, and admits a Cholesky decomposition.

Remark 4.1. An extension of the proposed method to the case when Ω is absorbing and an error analysis of the method will appear in a forthcoming paper.

REFERENCES

1. Aziz, A.K. and Kellogg, R.B., Finite Element Analysis of a Scattering Problem, Math. Comp. (to appear).

2. Aziz, A.K., Dorr, M.R. and Kellogg, R.B., Calculation of Electromagnetic Scattering by a Perfect Conductor, J. Computational Phys. (to appear).

3. Aziz, A.K., Dorr, M.R., and Kellogg, R.B., A New Approximation Method for an Exterior Problem for the Helmholtz Equation, (to appear).

4. Babuška, I. and Aziz, A.K., Survey Lectures on the Mathematical Foundations of the Finite Element Method, The Mathematical Foundations of the Finite Element Method with Applications to Partial Differential Equations, A.K. Aziz (editor), Academic Press, New York, 1972.

5. Bayliss, A., Gunzberger, M. and Turkel, E., Boundary Conditions for the Numerical Solution of Elliptic Equations in Exterior Regions, ICASE Report 80-1 (1978).

6. Bendali, A., Problemè aux limites exterieur et interieur pour le systѐme de Maxwell en regime harmonique, Repport intern N°: 59, Centre de Mathematiques Appliquées, Ecole Polytechnique, Paris, France.

7. Brezzi, F. and Johnson, C., On the Coupling of the Boundary Conditions and the Finite Element Method, Dept. of Computer Science, Chalmers University of Technology, Report 77.15R, 1977.

8. Grisvard, P., Caractérisation de quelques espaces d'interpolation, Arch. Rational Mech. Anal. 25 (1967), pp.40-30.

9. Giroire, J., Integral Equation Methods for Exterior Problems for the Helmholtz, Rapport Interne N° 40, Centre de Mathémiques Appliquées, Ecole Polytechnique, Paris, France.

10. Goldstein, C., Numerical methods for Helmholtz Type Equation in Unbounded

Domains, BNL-2643, Brookhaven Laboratory, Brookhaven, N.Y., 1979.

11. Kleinman, R.E. and Roach, G.F., Boundary Integral for Three Dimensional Helmholtz Equations, SIAM Rev. 16 (1978), pp.214-236.

12. Kreigsmann, G.A. and Morawetz, C.S., Numerical Solutions of Exterior Problems with the Reduced Wave Equation, J. Computational Plys. 28 (1978), pp.181-197.

13. Kreigsmann, G.A., and Morawetz, C.S., Solving the Helmholtz Equation with Variable Index of Refraction: 1, SIAM J. Sci. Stat. Comp. 1 (1980), pp.371-385.

14. MacCamy, R.C. and Marin, S.P., A Finite Element Method for Exterior Interface Problems, Internat'l. J. of Math. and Math. Sci. 3 (1980), pp.311-350.

15. MacCamy, R.C., Variational Procedures for a Class of Exterior Interface Problems, J. Math. Anal. Appl. 78 (1980), pp.248-266.

16. Müller, C., Foundation of Mathematical Theory of Electromagnetic Waves, Springer-Verlag, New York, 1969.

17. Schatz, A., An Observation Concerning Ritz-Galerkin Methods with Indefinite Bilinear Forms, Math. Comp. 28 (1974), pp.959-962.

18. Stewart, G.W., Introduction to Matrix Computations, Academic Press, New York, 1973.

19. Stratton, J.A., Electromagnetic Theory, McGraw-Hill, New York, 1941.

20. Zienkiewicz, O.C., Kelley, D.W., and Bettess, P., The Coupling of the Finite Element Method and the Boundary Solution Procedures, Int. J. for Numerical Methods in Engineering 11 (1977), pp.355-375.

21. Waterman, P.C., Matrix Formulation of Electromagnetic Scattering, Proc. IEEE 53 (1975), pp.805-812.

22. Waterman, P.C., New Formulation of Acoustic Scattering, J. of the Acoustical Soc. 45 (1969), pp.1417-1429.

Contemporary Mathematics
Volume 11, 1982

Boundary Value Problems Associated with First Order Elliptic Systems in the Plane

by

H. Begehr
Free University, Berlin

and

R. P. Gilbert*
University of Delaware

I. Introduction.

In this work we investigate the solvability of certain boundary value problems associated with generalized hyperanalytic functions. The generalized hyperanalytic functions extend the family of hyperanalytic functions much in the same way as the pseudoanalytic or generalized analytic functions extend the concept of analyticity. The hyperanalytic functions are hypercomplex valued functions which satisfy the elliptic differential system

$$Dw = 0, \text{ where } D := \frac{\partial}{\partial \bar{z}} + q(z)\frac{\partial}{\partial z} \tag{1}$$

Here $w := \sum_{k=0}^{r-1} e^k w_k(x,y)$, and $q(z) = \sum_{k=1}^{r-1} e^k q_k(x,y)$, are hypercomplex valued, that is the functions w_k, q_k are complex valued and e is a nilpotent such that $e^r=0$. It is usually assumed, furthermore, that q is a Hölder-continuous function in some regular domain $\Omega\subset\mathbb{R}^2$.

---

* This research was supported in part by the National Science Foundation through NSF Grant No. MCS 78-02452.

Douglis [7] developed the function theory associated with (1) and Begehr, Gilbert, and Hile studied various problems associated with the generalized hyperanalytic system [2] [3] [4] [8] [9] [10] [11] [14]

$$Dw + \sum_{k,\ell=0}^{n-1} e^k (A_{k\ell} w_\ell + B_{k\ell} \bar{w}_\ell) = F. \tag{2}$$

Contributions also have been made by Habetha [13], Kühn [15], and Wendland [12] [19]. It should be pointed out that the systems (2) are actually special cases of the "normalized" elliptic systems† investigated by Bojarski [5].

The results obtained in the case of Equation (2) and, in particular, for the special case where we are able to remain within the algebra of hypercomplex quantities, namely when we have the equation

$$Dw = aw + b\bar{w} = F, \tag{3}$$

with hypercomplex a, b, and F resemble closely those from the classical theory of Vekua [18] and Bers [1]. This leads to elegant constructive methods for solving the various boundary value problems such as the Hilbert and Riemann problems. One such approach from the classical theory, which was exploited extensively by Vekua [18], was to reduce various problems to related ones involving analytic functions. This required a Carleman type

---

† Actually no normal form exists for the case of 2r first-order elliptic systems in the plane. The reader is directed to the book by Wendland [19] for a discussion.

representation for the generalized analytic functions. In this paper we obtain an appropriate analogue to the Carleman representation for generalized hyperanalytic functions which permits a reduction to the case of hyperanalytic functions. We then use this to study the solvability of the Riemann-Hilbert problem.

II. Solvability of Riemann Hilbert Problems.

We investigate the so-called Hilbert boundary value problem associated with the hypercomplex system

$$(A)\quad \begin{cases} L[w] := Dw + \bar{A}w + B\bar{w} = F \text{ in } G, \\ \operatorname{Re}(\bar{\lambda}w) = \gamma \text{ on } \dot{G}. \end{cases} \qquad (1)$$

We refer to this as Boundary Value Problem (A). It is assumed, morover, that $\lambda \in C^1(\Gamma)$, $\Gamma := \dot{G}$, $\lambda\bar{\lambda} = 1$. The adjoint, homogeneous problem to (A) we refer to as $(\overset{\circ}{A}')$; it is given by

$$(\overset{\circ}{A}'):\quad \begin{cases} L'[w'] := Dw' - \bar{A}w' - B^{*}\bar{w}' = 0 \text{ in } G, \\ \operatorname{Re}(\lambda \frac{dt(z)}{ds} w') = 0 \text{ on } \dot{G}, \end{cases} \qquad (2)$$

where $B^{*} := \bar{B}\,\overline{t(z)}\,[t(z)]^{-1}$, $\frac{dt}{ds} := \frac{\partial t}{\partial z}\frac{dz}{ds} + \frac{\partial t}{\partial \bar{z}}\frac{d\bar{z}}{ds}$, and ds is the arc length differential. From Green's identity for hypercomplex quantities we have

$$\operatorname{Re}\{\frac{1}{i}\int_\Gamma ww'\,dt(z) - \int_G t_z(w'L[w] - wL'[w']\,dxdy\} = 0.$$

For $L[w]=F$ and $L'[w']=0$ this becomes

$$\operatorname{Re}\{\frac{1}{i}\int_\Gamma ww'\,dt(z) - \int_G t_z w' F\,dxdy\} = 0,$$

and for w' a solution of $(\overset{\circ}{A}')$ we obtain

$$\frac{1}{i}\int_\Gamma \gamma\lambda w'(z)\,dt(z) - \operatorname{Re}\int_G t_z\, w'(z) F(z)\,dxdy = 0. \qquad (3)$$

The solutions to Problem $(\overset{\circ}{A}')$ may be represented by means of the fundamental kernels in terms of a real, hypercomplex density, $f(z)$, as

$$w'(z) = -\frac{1}{2\pi i}\int_\Gamma \left(\frac{\Omega^{(1)}(\zeta,z)}{\lambda(\zeta)} + \frac{\overline{\Omega^{(2)}(\zeta,z)}}{\overline{\lambda(\zeta)}}\right) f(\zeta)\,ds.$$

This follows from the theory of singular integral equations. To establish (3) we compute using the Plemelj formula that $f(z)$ must satisfy

$$\int_{\Gamma} K_1(\zeta,z) f(\zeta)\, ds_{\zeta} = 0 \quad \text{where } z,\zeta=\zeta(s)\in\Gamma , \tag{4}$$

and

$$K_1(\zeta,z) =: -\frac{\text{Re}}{2\pi}\left\{\overline{\frac{dt\lambda(z)}{ds}\,(\Omega^{(1)}(\zeta,z)\lambda(\zeta)+\lambda(\zeta)\Omega^{(2)}(\zeta,z))}\right\} \tag{5}$$

The integral in (4) is to be taken in the Cauchy principal value sense. If we denote this equation in operator form as $\underset{\sim}{K}f=0$, and its adjoint as $\underset{\sim}{K}'\Phi=0$, then it may be easily demonstrated that the index of (4) is $\kappa=k-k'=0$. Here k and k' are the dimensions of the null spaces of $\underset{\sim}{K}$ and $\underset{\sim}{K}'$ respectively. If $\{f_1,f_2,\ldots,f_k\}$ are a complete system of solutions of (4) putting each of these into (3) generates solutions of $(\overset{o}{A}{}')$. However, it is possible that some of these correspond to the trivial solution, which occurs when $[\lambda\frac{dt}{ds}]^{-1}f$ takes on the boundary values of a hyperanalytic function Φ_j on each component of the boundary contours Γ_j which is, moreover, hyperanalytic in the domain G_j bounded by the closed contour Γ_j. If we designate $\{f_1,\ldots,f_{\ell'}\}$ as those densities to which linearly independent solutions (see [2]) $\{w_1',\ldots,w_{\ell'}'\}$ correspond, then the remaining densities $\{f_{\ell+1},\ldots,f_k\}$ satisfy boundary conditions of the form

$$f(z) = i\,\lambda(z)\,\frac{dt}{ds}\,\Phi^-(z) \text{ on } \Gamma . \tag{6}$$

Here the $\Phi^-(z)$ are meant to be hyperanalytic outside of $\overline{G}:=G+\Gamma$, and $\Phi(\infty)=0$. Hence the hyperanalytic functions satisfy the homogeneous boundary conditions

$$\overset{o}{A}{}'_*: \text{Re}[\Phi^-\lambda(z)\,\frac{dt}{ds}] = 0 \text{ on } \Gamma. \tag{7}$$

Vekua refers to problems of this type as being concomitant to $(\overset{\circ}{A}{}')$, and denotes it as $(\overset{\circ}{A}{}'_*)$. If ℓ'_* is the number of linearly independent solutions to $(\overset{\circ}{A}{}'_*)$ then clearly $k=\ell'+\ell'_*$.

We return to a discussion of Problem (A) where we assume that f=0 in what follows. The solutions to this problem may be expressed in terms of generalized Cauchy kernels as

$$w(z)=w_1(z)+W_2(z)=\underset{\sim}{C}[\gamma/\overline{\lambda}](z)+\underset{\sim}{C}[i\mu/\overline{\lambda}](z),$$

$$\text{where}\quad \underset{\sim}{C}[\phi]:=\frac{1}{2\pi i}\int_\Gamma \Omega^{(1)}(z,\zeta)\phi(\zeta)\,dt(\zeta)-\Omega^{(2)}(z,\zeta)\,\overline{\phi(\zeta)}\,\,\overline{dt(\zeta)}. \qquad (8)$$

From the Plemelj relations it is seen that the density μ must satisfy the integral equation

$$\gamma_o(\zeta) = \int_\Gamma K_1(\zeta,z)\mu(z)\,ds_z,$$

where (9)

$$\gamma_o(\zeta):=\gamma(\zeta)-\operatorname{Re}[\overline{\lambda(\zeta)}w_1^+(\zeta)] = -\operatorname{Re}[\overline{\lambda(\zeta)}w_1^-(\zeta)].$$

The Problem $(\overset{\circ}{A}_*)$ concomitant to $(\overset{\circ}{A})$ has the boundary conditions $\operatorname{Re}[\lambda^{-1}(z)\Phi^-(z)]=0$ on Γ where Φ is hyperanalytic outside $G+\Gamma$ and $\Phi(\infty)=0$. If, as before, ℓ and ℓ^* denote the linear independent solutions of $(\overset{\circ}{A})$ and $(\overset{\circ}{A}_*)$ we have $k=\ell+\ell_*$. In order that (9) be solvable it is necessary and sufficient that the nonhomogeneous data γ_o satisfy the auxiliary conditions

$$\int_\Gamma \gamma_o(\zeta)f_j(\zeta)\,ds_\zeta = 0 \quad (j=1,2,\ldots,k), \qquad (10)$$

where the f_j are the solutions to integral equation (4). These solutions may be broken up into two groups $\{f_1,\ldots,f_{\ell'}\}$ and $\{f_{\ell'+1},\ldots,f_k\}$ such that $f_j=i\lambda(z)\frac{dt(z(s))}{ds}w'_j(z)$ for $(j=1,\ldots,\ell')$, and $f_j=i\lambda(z)\frac{dt(z(s))}{ds}\,\Phi_j(z)$ for $(j=\ell'+1,\ldots,k)$ where $z\in\Gamma$. Here w'_j and Φ_j are solutions of the Problem $(\overset{\circ}{A}{}')$ and $(\overset{\circ}{A}{}'_*)$ respectively. The condition (10) for γ_o given by (9) becomes

$$-\int_\Gamma \gamma_o(\zeta) f_j(\zeta)\,ds = \int_\Gamma \frac{\gamma(\zeta) w_j'(\zeta)\,dt(\zeta)}{i\overline{\lambda(\zeta)}} + \mathrm{Re}[i\int_\Gamma w_j'(\zeta) w_1^+(\zeta)\,dt(\zeta)]$$

for $j=1,\ldots,\ell'$, whereas for $j=\ell'+1,\ldots,k$ we have

$$\int_\Gamma \gamma_o(\zeta) f_j(\zeta)\,ds = \mathrm{Re}[i\int_\Gamma w^-(\zeta)\ \Phi_j^-(\zeta)\,dt(\zeta)] = 0.$$

Consequently, the conditions (10) are seen to hold if (3) (with $F\equiv 0$) holds. From the above discussion one obtains a Fredholm type theorem for Problem (A); namely (A) is solvable if and only if (3) holds for w' an arbitrary solution of $(\overset{o}{A}{}')$.

We define the index of Problem (A) as

$$n := \frac{1}{2\pi}\int_\Gamma d\ \ell n\ \lambda = \frac{1}{2\pi i}\int_\Gamma d\ \ell n\ \lambda_o, \text{ where } \lambda_o = \text{c.p.}[\lambda], \tag{11}$$

which follows from the identity $\ell n\lambda := \ell n\lambda_o + \sum_{k=1}^{r-1} \frac{(-1)^{k-1}}{k}\left(\frac{\lambda-\lambda_o}{\lambda_o}\right)^k$, $\lambda_o \neq 0$. The index of Problem $(\overset{o}{A}{}')$ correspondingly is given by

$$n' := \frac{1}{2\pi i}\int_\Gamma d\ \ell n\ \overline{(\lambda(z)\frac{dt}{ds})}$$

$$= -\frac{1}{2\pi i}\int_\Gamma d\ \ell n\lambda(z) - \frac{1}{2\pi i}\int_\Gamma \ell n\frac{\partial t}{\partial z} - \frac{1}{2\pi i}\int_\Gamma d\ \ell n z'(s) \tag{12}$$

$$= -n + m-1,$$

where $m+1$ is the connectivity of G, since $\frac{\partial t}{\partial \bar z}$ is a nilpotent.

We investigate next the possibility of developing a Vekua-type canonical form for Problem (A). Let us decompose the index given in (11) in terms of the boundary components Γ_j as

$$n = \frac{1}{2\pi i}\sum_{i=0}^{m}\int_{\Gamma_i} d\ \ell n\lambda = \sum_{i=0}^{m} n_i. \tag{13}$$

Following Vekua [18] we introduce the functions

$$\Omega_n(z) := \begin{cases} \prod_{i=1}^{n} [t(z)-t(a_i)] , & n>0, \\ 1 , & n=0, \\ \prod_{i=1}^{-n} [t(z)-t(a_i)]^{-1}, & n<0, \end{cases} \tag{14}$$

and

$$\hat{\lambda}(z) := \lambda(z)\overline{\Omega_n(z)} \prod_{k=1}^{m} [\overline{t(z)-t(z_k)}]^{-n_k} \exp(\overline{-\underset{\sim}{T}A}), \tag{15}$$

$$\text{where } \underset{\sim}{T}A := -\frac{1}{\pi} \int_G \frac{t_\zeta(\zeta)A(\zeta)}{t(\zeta)-t(z)} \, d\xi d\eta . \tag{16}$$

We notice that if $\lambda(z)\neq 0$ on Γ, then $\hat{\lambda}(z)\neq 0$ on Γ. If $\Gamma := \sum_{i=0}^{m} \Gamma_i$, where Γ_0 is the outer boundary then

$$\int_{\Gamma_0} d\,\ell n\,\hat{\lambda} = \int_{\Gamma_o} d\,\ell n\,\lambda + \int_{\Gamma_o} d\,\ell n\,\overline{\Omega_n} + \sum_{k=1}^{m} n_k \int_{\Gamma_0} d\,\ell n(t(z)-t(z_k))$$

$$= 2\pi \, (n_0 - n + \sum_{k=1}^{m} n_k) = 0, \tag{17}$$

and

$$\int_{\Gamma_k} d\,\ell n\,\hat{\lambda} = \int_{\Gamma_k} d\,\ell n\,\lambda + n_k \int_{\Gamma_k} d\,\ell n(t(z)-t(z_k)) = 2\pi(n_k - n_k) = 0 \tag{18}$$

Here $\hat{\lambda} = \hat{\rho}\exp(i\hat{\sigma})$, and $\lambda = \rho \exp(i\sigma)$ where $\hat{\rho}$ is hyperreal and $\hat{\sigma}$ is single-valued hyperreal. Indeed,

$$\hat{\sigma} := \arg\lambda - \arg\Omega_n + \sum_{k=1}^{m} n_k \arg[t(z)-t(z_k)] + \operatorname{Im} \underset{\sim}{T}A.$$

Moreover, if $\lambda\in C^{\nu}(\Gamma)$, then as $A\in L_p(G+\Gamma)$ one has that $\underset{\sim}{T}A\in C^{\alpha}(\Gamma)$, with $\alpha = \frac{p-2}{p}$, and from this it follows that $\hat{\sigma}\in C^{\tau}(\Gamma)$, where $\tau = \min\{\nu,\alpha\}$.

<u>Claim</u>: $\hat{\sigma}$ <u>may be represented on</u> Γ <u>in the form</u> $\hat{\sigma}=\rho - i\sigma_* + \pi\alpha$, <u>where</u> $p(z)$ <u>is hyperanalytic in</u> G <u>and Hölder-continuous in</u> $G+\Gamma, G=\mathrm{Im}\rho$. <u>Moreover</u>, $\alpha(z)$ <u>is a piecewise constant function on</u> Γ, <u>that is</u>, $\alpha=0$ <u>of</u> Γ_0 <u>and</u> $\alpha=\alpha_k$ <u>on</u> Γ_k $(k=1,\ldots,m)$, <u>where the</u> α_k <u>and</u> $p(z)$ <u>are uniquely determined from</u> $\hat{\sigma}$.

<u>Proof</u>: It is of interest to first investigate the linear space of solutions to the hyperanalytic Problem (D)

$$(D): \quad D\Phi = 0 \text{ in } G, \ \mathrm{Re}\Phi = \gamma \text{ on } \Gamma. \tag{19}$$

We denote the adjoint, homogeneous problem as $(\overset{o}{D}{}')$

$$(\overset{o}{D}{}'): \ D\Phi' = 0 \text{ in } G, \ \mathrm{Re}\left(\frac{dt}{ds}\Phi'\right) = 0 \text{ on } \Gamma. \tag{20}$$

If $\Phi'(z) = \sum_{\ell=0}^{r-1} e^{\ell}\phi'_{\ell}(z)$, then the components satisfy the equations

$$\frac{\partial\phi'_o}{\partial\bar{z}} = 0, \ \frac{\partial\phi'_{\ell}}{\partial\bar{z}} = -\sum_{p=1}^{\ell} q_p \frac{\partial\phi^{(k)}_{\ell-p}}{\partial z}, \ (\ell=1,\ldots,r-1) \text{ in } G.$$

Let $u_k(x,y)$ be the harmonic measure of Γ_k in G; then $u_k = 0$ on Γ_0 and $u_k = \delta_{kj}$ on Γ_j for $1 \le k,j \le m$. The harmonic measures may, moreover be represented in terms of the Green's function as

$$u_k(x,y) = \int_{\Gamma_k} \frac{\partial g}{\partial n_t}(t,z)\,ds_t \quad (k=1,\ldots,m). \tag{21}$$

It is now possible to identify the complex component of the Problem $(\overset{o}{D}{}')$ with a u_k $(k=1,\ldots,m)$. This is seen by noting that

$$\frac{du_k}{ds} = \frac{\partial u_k}{\partial z}\frac{dz}{ds} + \frac{\partial u_k}{\partial\bar{z}}\frac{d\bar{z}}{ds} = 2\,\mathrm{Re}\left(\frac{\partial u_k}{\partial z}\frac{dz}{ds}\right) = 0 \text{ on } \Gamma,$$

and that $\frac{\partial t}{\partial z} = 1 + \sum_{\ell=1}^{r-1} e^{\ell} \frac{\partial t_\ell}{\partial z}$, whereas $\frac{\partial t}{\partial \bar{z}}$ is nilpotent. Since the boundary conditions of (20) may be written componentwise as

$$\mathrm{Re}(z'(s)\phi_o(z)) = 0 \quad \text{on } \Gamma, \tag{22}$$

$$\mathrm{Re}(z'(s)\phi_\ell(z)) = -\mathrm{Re}\Big[\sum_{k=0}^{\ell-1}\Big(\frac{\partial t_{\ell-k}}{\partial z}\frac{dz}{ds} + \frac{\partial t_{\ell-k}}{\partial \bar{z}}\frac{d\bar{z}}{ds}\Big)\phi_k\Big] \text{ on } \Gamma.$$

There are m linearly independent solutions of the problem $\frac{\partial \phi_o}{\partial \bar{z}} = 0$ in G, $\mathrm{Re}(z'(s)\phi_o) = 0$ on Γ, and these may be given explicitly as $\phi_o^{(k)}(z) := \frac{\partial u_k}{\partial z}$, (k=1,2,...m). The nilpotent components $\phi_\ell^{(k)}(z)$, (ℓ=1,...,r-1) are then given explicitly in term of the Green's function G^I and the Neumann function G^{II} as

$$\phi_\ell^{(k)}(z) = \theta_\ell^{(k)}(z) + i\sum_{p=1}^{\ell}\int_G \{q_p \frac{\partial \phi_{\ell-p}^{(k)}}{\partial \zeta}[G_\zeta^I + G_\zeta^{II}] + \overline{q_p}\,\overline{\frac{\partial \phi_{\ell-p}^{(k)}}{\partial \zeta}}[G_{\bar{\zeta}}^I - G_{\bar{\zeta}}^{II}]\, d\zeta\wedge d\bar{\zeta} \tag{23}$$

where

$$\phi_\ell^{(k)}(z) := -e^{\theta(z)}\int_\Gamma \mathrm{Re}\sum_{\nu=0}^{\ell-1}\Big[\frac{\partial t_{\ell-\nu}}{\partial \zeta}\frac{d\zeta}{ds} + \frac{\partial t_{\ell-\nu}}{\partial \bar{\zeta}}\frac{d\bar{\zeta}}{ds}\Big]\phi_\nu(\zeta).$$

$$\cdot e^{-\mathrm{Re}\,\theta(\zeta)}[d_n G^I - i dG^{II}] + iC_\ell^{(k)},$$

$$\text{where} \quad \theta(z) := i\int_\Gamma \arg\frac{d\zeta}{ds}\,[d_n G^I - i dG^{II}],$$

and the $C_\ell^{(k)}$ are constants which may be fixed arbitrarily. The above method allows us to determine m linearly independent solutions $\cdot\Phi^{(k)}$ to Problem $(\overset{o}{D}')$.

Next we turn to Problem (D), and note that solvability demands that condition (3) be imposed, namely

$$\frac{1}{i}\int_\Gamma \gamma(z)\Phi^{(k)}(z)\,ds_z = 0. \tag{24}$$

Considering just the complex part yields

$$\frac{1}{i}\int_{\Gamma} \gamma_o(z)\phi_o^{(k)}(z)\,dz = \int_{\Gamma} \gamma_o(z)\frac{\partial u_k}{\partial n}\,ds = 0,(k=1,2,\ldots,m). \quad (25)$$

Assuming that $\gamma_o(z)$ may be written in the form $\gamma_o(z) = \tilde{\gamma}_o(z)+c_{o,j}$ on Γ_j where the $C_{o,j}$ are real contants and $\tilde{\gamma}_o$ is continuous on Γ we obtain the following system to determine the $c_{o,j}$,

$$\sum_{j=1}^{m} A_{kj}\, c_{o,j} = -\int_{\Gamma}\tilde{\gamma}_o \frac{\partial u_k}{\partial n}ds \quad (k=1,\ldots,m), \quad (26)$$

where the $A_{kj} = \int_{\Gamma_j} \frac{\partial u_k}{\partial n}\,ds = \int_{\Gamma} u_j \frac{\partial u_k}{\partial n}\,ds = \frac{1}{i}\int u_j\phi_o^{(k)}\,dz$

$$= \int_G \bar{\phi}_o^{(j)}\,\phi_o^{(k)}\,dxdy. \quad (27)$$

Since the determinant of the system is a Gram determinant for the linearly independent system $\{\phi_o^{(1)},\ldots,\phi_o^{(m)}\}$ the system has a unique solution. Vekua has shown that the $c_{o,k}$ may be written as

$$c_{o,k} = i\int_{\Gamma} \tilde{\gamma}_o\phi_o^{(k)*}dz, \quad (28)$$

where $\phi_o^{(k)*} = 2\frac{\partial u_k^*}{\partial z}$, and $u_k^* = a_{k_1}\, u_1+\ldots+a_{km}\, u_m$. (29)

The a_{kj} may be determined from the identity [18] (pg. 263)

$$\sum_{i=1}^{m} A_{ki}\, a_{ij} = \sum_{i=1}^{m} A_{ik}\, a_{ij} = \delta_{kj}\,. \quad (30)$$

This shows that we may always solve the boundary value problems associated with the first component of (D), namely one may rewrite the problem, by asking that $\mathrm{Re}\phi_o = \tilde{\gamma}_o(z)+c_o(z)$ on Γ where $\tilde{\gamma}_o := \gamma_o - c_o$, such that $c(z) = 0$ on Γ_o and $c(z) = c_k$ on Γ_k $(k=1,\dots,m)$. As $\tilde{\gamma}_o$ satisfies condition (25) for this choice of $c(z)$ we see that Problem (D_o) always has a solution.

We turn next to the $\ell^{\underline{th}}$ component of Problem (D),

$$(\tilde{D}_\ell): \begin{cases} \dfrac{\partial \phi_\ell}{\partial \bar{z}} = -\displaystyle\sum_{p=1}^{\ell} q_p \dfrac{\partial \phi_{\ell-p}}{\partial z} \quad \text{in } G, \\[2ex] \mathrm{Re}(\phi_\ell) = \gamma_\ell \quad \text{on} \quad \Gamma, \; (\ell=1,2,\dots,r-1). \end{cases} \tag{32}$$

The adjoint homogeneous problem to $(\tilde{D}_\ell)$ is

$$(\overset{o}{D}{}'_\ell): \left\{ \dfrac{\partial \phi'_\ell}{\partial \bar{z}} = 0 \text{ in } G, \; \mathrm{Re}(z'(s)\phi'_\ell(z)) = 0 \text{ on } \Gamma. \right.$$

Since we are merely interested in the question of solvability of $(\tilde{D}_\ell)$, it suffices to notice that we reduce $(\tilde{D}_\ell)$ to a homogeneous Cauchy Riemann equation by introducing a particular solution. This reduction, in turn, changes the data γ_ℓ on Γ; hence, we need only consider the problem

$$(D_\ell): \left\{ \dfrac{\partial \phi_\ell}{\partial \bar{z}} = 0 \text{ in } G, \; \mathrm{Re}\phi_\ell = \tilde{\gamma}_\ell \text{ on } \Gamma \; (\ell=1,\dots,m). \right. \tag{33}$$

The proof that (D_ℓ) is solvable proceeds exactly as in the case of (D_o). Hence we may always solve the Problem (D). Indeed, if we rewrite the hypercomplex data in (19) as $\gamma(z) = \tilde{\gamma}(z) + c(z)$ where $c(z)$ a piecewise, hyperreal constant on Γ such that $c(z) = 0$ on Γ_o and

$$c(z) = \sum_{\ell=0}^{r-1} e^\ell c_{k\ell} \quad \text{on} \quad \Gamma_k \; (k=1,2,\dots m), \tag{34}$$

then there exists a hyperanalytic function $\Psi(z)$ in G satisfying the boundary conditions $\mathrm{Re}\Psi = \tilde{\gamma}$. This implies that

$$\gamma(z) = \Psi(z) - i\gamma_*(z) + c(z) \quad \text{on } \Gamma \tag{35}$$

where $\gamma_*(z) = \mathrm{Im}(\Psi(z))$. [It appears that the hypercomplex solvability conditions (24) might be avoided altogether and just the components worked with!]

Returning to equations (15) (16) we introduce the hypercomplex polar representation for the data, namely

$$\overline{\lambda} = \hat{\rho}\ \exp\{-i(p-i\sigma_*+\pi\alpha)\}\ \Omega_n^{-1} \prod_{k=1}^{m} [t(z)-t(z_k)]^{n_k} \exp \underset{\sim}{T}A,$$

where $\hat{\rho} := \rho\ |\Omega_n| \prod_{k=1}^{m} |t(z)-t(z_k)|^{-n_k} \exp\{-\mathrm{Re}\underset{\sim}{T}A\}$,

and $|\Omega_n|$, and $|t(z)-t(z_n)|$ are the radial coordinates of Ω_n and $t(z)-t(z_n)$ respectively. Consequently by introducing the new unknown

$$\tilde{w}(z) := w\ \exp(ip(z)) \prod_{k=1}^{m} [t(z)-t(z_k)]^{n_k} \exp(\underset{\sim}{T}A) \tag{36}$$

the boundary condition of Problem (A) may be rewritten as

$$\mathrm{Re}\{\overline{\Omega}_n \exp(-i\pi\alpha(z))\tilde{w}(z)\} = \tilde{\gamma} \text{ on } \Gamma \tag{37}$$

where

$$\tilde{\gamma} := \frac{\gamma}{\rho}\ |\Omega_n| \prod_{k=1}^{m} |t(z)-t(z_k)|^{n_k} \exp\{\mathrm{Re}\underset{\sim}{T}A+\sigma_*(z)\}$$

The differential equation of (A) reduces to the form

$$D\tilde{w} + \tilde{B}\ \overline{\tilde{w}} = \tilde{F}, \text{ where}$$

$$\tilde{B} := B\ \exp\{-2i[\mathrm{Re}\ p(z) - \mathrm{Im}\underset{\sim}{T}A - \sum_{k=1}^{m} n_k \arg(t(z)-t(z_k))]\}, \tag{38}$$

and

$$\tilde{F} := F \exp(-ip(z)) \prod_{k=1}^{m} [t(z)-t(z_k)]^{n_k} \exp(\underset{\sim}{TA}).$$

We refer to the reduced problem as the cannonical form of (A)

$$(C): \begin{cases} D\tilde{w} + \tilde{B}\overline{\tilde{w}} = \tilde{F} \text{ in } G, \\ \\ \mathrm{Re}\{\overline{\Omega}_n \exp(-i\pi\alpha)\tilde{w}\} = \tilde{\gamma} \text{ on } \Gamma. \end{cases}$$

The homogeneous adjoint problem to (C) is

$$(\overset{o}{C}'): \begin{cases} D\tilde{w}' - \tilde{B}^{*}\, \overline{\tilde{w}}' = 0 \text{ in } G, \\ \\ \mathrm{Re}\{\Omega \exp(i\pi\alpha)\, \dfrac{dt(z(s))}{ds}\, \tilde{w}'\} = 0 \text{ on } \Gamma, \end{cases}$$

where $\tilde{B}^{*} := \overline{t_z}\ (t_z)^{-1}\, \overline{\tilde{B}}$.

III. Carleman-Type Theorems

It is convenient for further development if something equivalent to the Similarity Principal for generalized analytic functions would be true for our case. By this we mean that it would be possible to represent all solutions as a product, or sum of products, of analytic (hyperanalytic) functions and a nowhere vanishing C^{α}-function. To this end, we consider constructing solutions of the Douglis system, which we now write in matrix form as

$$D\underset{\sim}{w} + A\underset{\sim}{w} + B\overline{\underset{\sim}{w}} = 0, \tag{1}$$

where

$$D := \frac{\partial}{\partial \bar{z}} + Q\frac{\partial}{\partial z}, \quad Q := \begin{bmatrix} 0 & & & \\ q_1 & \ddots & & 0 \\ \vdots & \ddots & \ddots & \\ \vdots & & \ddots & \ddots \\ q_{r-1} & \dots & q_1 & 0 \end{bmatrix},$$

$$A: \begin{bmatrix} a_{00} & & & 0 \\ \vdots & \ddots & & \\ \vdots & & \ddots & \\ a_{r-1,0} & \dots & & a_{r-1,r-1} \end{bmatrix}, \quad B := \begin{bmatrix} b_{00} & & & 0 \\ \vdots & \ddots & & \\ \vdots & & \ddots & \\ b_{r-1,0} & \dots & & b_{r-1,r-1} \end{bmatrix}, \tag{2}$$

$\underset{\sim}{w} := [w_0, \dots, w_{r-1}]'$. It is convenient for us to define the matrix†

$$C := A + B\begin{bmatrix} \frac{\overline{w}_0}{w_0} & & \\ & \ddots & \\ & & \frac{\overline{w}_{r-1}}{w_{r-1}} \end{bmatrix} = \begin{bmatrix} a_{00} + b_{00}\frac{\overline{w}_0}{w_0} & & & \\ \vdots & \ddots & & \\ \vdots & & \ddots & \\ a_{r-1,0} + b_{r-1,0}\frac{\overline{w}_0}{w_0}, & \dots, & & a_{r-1,r-1} + b_{r-1,r-1}\frac{\overline{w}_{r-1}}{w_{r-1}} \end{bmatrix}$$

† In the event that $w_\nu = 0$ we replace $\overline{w}_\nu / w_\nu$ by 1 in the definition of C.

$$= \begin{bmatrix} c_{00} & & & \\ \cdot & \ddots & & 0 \\ \cdot & & \ddots & \\ \cdot & & & \ddots \\ c_{r-1,0} & \cdots & & c_{r-1,r-1} \end{bmatrix}. \tag{3}$$

We notice, moreover, that $Cw=Aw+B\overline{w}$; hence, the system (1) may be written as $Dw + Cw = 0$. We consider next the matrix differential equation

$$DT + CT = 0, \tag{4}$$

with the asymptotic condition $T(\infty)=E$ (the identity matrix). The $k^{\underline{th}}$ column $T^{(k)}$ of the matrix T may be associated component for component with the hypercomplex function $\tau^{(k)} = \sum_{\nu=0}^{r-1} \tau_\nu^{(k)} e^\nu$ which satisfies the integral equation

$$\tau^{(k)} = \frac{1}{\pi} \int_D \frac{t_\zeta(\zeta)}{t(\zeta)-t(z)} \left\{ \sum_{\nu=0}^{r-1} \sum_{\mu=0}^{\nu} C_{\nu\mu} \tau_\mu^{(k)} e^\nu \right\} d\xi d\eta + e^k. \tag{5}$$

Apply the (hypercomplex) operator D to both sides of (5) yields

$$\left(\frac{\partial}{\partial \overline{z}} + \sum_{\nu=1}^{r-1} q_\nu e^\nu \frac{\partial}{\partial z} \right) \tau^{(k)} = - \sum_{\nu=0}^{r-1} \sum_{\mu=0}^{\nu} C_{\nu\mu} \tau_\mu^{(k)} e^\nu,$$

so

$$\frac{\partial}{\partial \overline{z}} \tau_\nu^{(k)} + \sum_{\ell=0}^{r-1} q_{\nu-\ell} \frac{\partial}{\partial z} \tau_\ell^{(k)} + \sum_{\ell=0}^{\nu} C_{\nu\ell} \tau_\ell^{(k)} = 0, \quad (0 \le \nu \le r-1). \tag{6}$$

The bounded solutions of (5) are known to be generalized hypercomplex constants ([9] pg. 18), and hence $\tau^{(k)}$ must have the form

$$\tau^{(k)} = e^k \exp \omega^{(k)} \tag{7}$$

where $\omega^{(k)} \in C^\alpha(\mathbb{C})$, $\alpha = \frac{p-2}{p}$, $p>2$. Consequently, $\tau_\nu^{(k)} = 0$ for $0 \le \nu \le k-1$, but $\tau_k^{(k)}(z) \neq 0$ in $\mathbb{C}$. From this we see that the components $\tau_\nu^{(k)}$ of the vector $T^{(k)}$ fulfill

$$\frac{\partial}{\partial \bar{z}} \tau_{\nu}^{(k)} + \sum_{\ell=k}^{\nu-1} q_{\nu-\ell} \frac{\partial}{\partial z} \tau_{\ell}^{(k)} + \sum_{\ell=k}^{\nu} C_{\nu\ell} \tau_{\ell}^{(k)} = 0 \quad (0 \leq k \leq \nu \leq r-1),$$

or

$$\frac{\partial}{\partial \bar{z}} T^{(k)} + Q \frac{\partial}{\partial z} T^{(k)} + C\, T^{(k)} = 0, \tag{8}$$

where

$$T = \begin{bmatrix} T_0^{(0)} & & 0 \\ \vdots & \ddots & \\ \tau_{r-1}^{(0)} & \cdots & \tau_{r-1}^{(r-1)} \end{bmatrix}, \ \det T = \prod_{\nu=0}^{r-1} \tau_{\nu}^{(\nu)} \neq 0 \text{ in } \hat{\mathbb{C}}. \tag{9}$$

Therefore T^{-1} exists and from $T^{-1}T=E$ one easily verifies the identity

$$D^{*}T^{-1} = -T^{-1}[DT]T^{-1}, \tag{10}$$

where

$$D^{*} := \frac{\partial}{\partial \bar{z}} + T^{-1}\, QT \frac{\partial}{\partial z}.$$

If we introduce the function $\phi := T^{-1}w$, then

$$D^{*}\phi = (D^{*}T^{-1})w + T^{-1}Dw = T^{-1}[Dw - (DT)T^{-1}w]$$

$$= T^{-1}[Dw + CTT^{-1}w] = T^{-1}[Dw + Aw + B\bar{w}] = 0.$$

However, $T^{-1}QT$ is in general not quasidiagonal, but only nilpotent so we cannot guarantee that ϕ is hyperanalytic

Remark: In the Pascali systems [16] we have $Q=0$. Then if $w=T\phi$ as T satisfies $\frac{\partial T}{\partial \bar{z}} + CT = 0$ we have $\frac{\partial \phi}{\partial \bar{z}} = 0$, that is ϕ is an analytic vector.

Remark: We notice that when $B=0$ the matrix C is quasidiagonal; consequently Q^{*} is in this instance quasidiagonal.

If we identify the vector w with the hypercomplex quantity $w := \sum_{\nu=0}^{r-1} w_{\nu}(z)e^{\nu}$, where $w_{\nu}(z) := \sum_{k=0}^{\nu} \tau_{\nu}^{(k)}\phi_k = \sum_{k=0}^{r-1} \tau_{\nu}^{(k)}\phi_k$, then

$$w = \sum_{\nu=0}^{r-1} \sum_{k=0}^{r-1} \tau_\nu^{(k)} \phi_k e^\nu = \sum_{k=0}^{r-1} \sum_{\nu=k}^{r-1} \tau_\nu^{(k)} e^\nu \phi_k = \sum_{k=0}^{r-1} \phi_k \tau^{(k)}, \tag{11}$$

where

$$\tau^{(k)} := \sum_{\nu=0}^{r-1} \tau_\nu^{(k)} e^\nu = e^k \exp\omega^{(k)}. \tag{12}$$

The representation (11), (12) for solutions to equation (1) with functions ϕ satisfying $D^*\phi=0$ is a version of the Carleman theorem. We investigate next the local behavior around the zeros of $w(z)$. Since, in general

$$Q^* := T^{-1}QT = \begin{bmatrix} 0 & & & & \\ q_{10} & \cdot & & 0 & \\ \cdot & & \cdot & & \\ \cdot & & & \cdot & \\ \cdot & & & & \cdot \\ q_{r-1,0} & & & q_{r-1,r-2} & 0 \end{bmatrix} \tag{13}$$

the components of $D^*\phi = 0$ satisfy

$$\frac{\partial}{\partial \bar{z}} \phi_k = 0, \quad \frac{\partial}{\partial \bar{z}} \phi_\nu + \sum_{\mu=k}^{\nu-1} q_{\nu\mu} \frac{\partial}{\partial z} \phi_\mu = 0 \quad (k<\nu<r-1) \tag{14}$$

where $\phi_k(z)$ is the first non identically vanishing component of ϕ. Since $\phi_k(z)$ is analytic the zeros of ϕ are isolated. For simplicity, in what follows, we shall assume $\phi_o(z)\neq 0$, but that $\phi_o(z_o)=0$ and z_o is taken without loss of generality to be the origin. It shall, moreover, be assumed that $q_{\nu\mu} \in C^\alpha(D)$ for some α $(0<\alpha<1)$. Since $\phi_o(z)=0$ then in some $N(0)$ we have $\phi_0(z)=z^{n_o}\phi_{oo}(z)$ where $n_o \in \mathbb{N}$, and $\phi_{oo}(z)$ is analytic in $N(0)$, and $\phi_{oo}(0)\neq 0$.

<u>Lemma: In a neighborhood of the origin</u> $N(0)$ $\phi_\mu(z)$ <u>has the representation</u>

$$\phi_\mu(z) = z^{n_\mu}\phi_{\mu\mu}(z) + \sum_{\nu=0}^{\mu-1} \{\phi_{\mu\nu}(z) + p_{\mu\nu}(\frac{\bar{z}}{z}) \frac{\bar{z}}{z}\} z^{n_\nu}, \tag{15}$$

($\mu=0,1,\ldots,r-1$) <u>where</u> $n_\nu \in \mathbb{N}$, $\phi_{\mu\nu} \in C^\alpha(N)$ <u>for</u> <u>some</u> $\alpha(0<\alpha<1)$, <u>and</u> $\phi_{\mu\nu} \in D_{1,p}$ $(2<p)$. <u>Moreover</u>, $\phi_{\mu\mu}$ <u>is</u> <u>analytic</u> <u>in</u> N, $\phi_{\mu\mu}(0)\neq 0$, <u>and</u> $p_{\mu\nu}$ is a polynomial and $\deg p_{\mu\nu} \leq \min\{\mu-\nu-1, n_\nu-1\}$.

<u>Proof</u>: In what follows we use $\underset{\sim}{T}$ to denote the <u>complex</u>, Pompieu operator

$$(\underset{\sim}{T}f)(z) := -\frac{1}{\pi}\int_N f(\zeta)\,\frac{d\xi d\eta}{\zeta - z}\,. \tag{16}$$

Using the Cauchy-Pompieu formula one obtains the identity

$$\frac{1}{\mu+1}\left(\frac{\bar{z}}{z}\right)^{\mu+1} = \chi(z) + \underset{\sim}{T}\left(\frac{\bar{z}}{z}\right)^{\mu}\frac{1}{z}\,, \tag{17}$$

where χ is analytic in $N(0)$. We prove (15) by induction. To this end let us assume that it is true for all $\mu, 0\leq\mu\leq k-1$, with $k<r$; then

$$\frac{\partial}{\partial\bar{z}}\phi_k = -\sum_{\mu=0}^{k-1} q_{k\mu}\frac{\partial}{\partial\bar{z}}\phi_\mu = -\sum_{\mu=0}^{k-1}\sum_{\nu=0}^{\mu}\Big[q_{k\mu}\frac{\partial}{\partial\bar{z}}\phi_{\mu\nu}$$

$$+ n_\nu\frac{q_{k\mu}\phi_{\mu\nu} - q_{k\mu}(0)\phi_{\mu\nu}(0)}{z}\Big]z^{n_\nu} - \sum_{\mu=0}^{k-1}\sum_{\nu=0}^{\mu-1}\Big[n_\nu q_{k\mu}(0)\phi_{\mu\nu}(0)$$

$$+ \hat{p}_{\mu\nu}\left(\frac{\bar{z}}{z}\right)\frac{\bar{z}}{z}\Big]z^{n_\nu-1} - \sum_{\mu=0}^{k-1} n_\mu q_{k\mu}(0)\phi_{\mu\mu}(0)z^{n_\mu-1} =$$

$$= \sum_{\nu=0}^{k-1}\Big[\tilde{\phi}_{k\nu}(z)z^{n_\nu} - \tilde{p}_{k\nu}\left(\frac{\bar{z}}{z}\right)z^{n_\nu-1}\Big]. \tag{18}$$

We note that as $f(z) := q_{k\mu}\ \phi_{\mu\nu} \in C^\alpha(N)$, $(0<\alpha<1)$ we have $[f(z)-f(0)]z^{-1} \in L_p(N)$ for $(2<p<\frac{2}{1-\alpha})$. Using this and the fact that $\frac{\partial}{\partial\bar{z}}\phi_{\mu\nu} \in L_p(N)$ we have $\tilde{\phi}_{k\nu} \in L_p(N)$. Finally, $\tilde{p}_{k\nu}$ is a polynomial such that $\deg \tilde{p}_{k\nu} \leq \min\{k-1-\nu, n_\nu\}$.

Using the Cauchy-Pompieu formula we then obtain

$$\phi_k(z) = \tilde{\phi}_k - \underset{\sim}{T}\left(\sum_{\nu=0}^{k-1} [\tilde{\phi}_{k\nu}(z) + \tilde{p}_{k\nu}(\tfrac{\bar{z}}{z})\tfrac{1}{z}] z^{n_\nu}\right)$$

where $\tilde{\phi}_k$ is analytic in it. From the identity

$$\frac{1}{\zeta - z} = \sum_{\lambda=0}^{n_\nu - 1} \frac{z^\lambda}{\zeta^{\lambda+1}} + \frac{z^{n_\nu}}{\zeta^{n_\nu}} \frac{1}{\zeta - z}$$

we calculate

$$-\underset{\sim}{T}(\tilde{\phi}_{k\nu}(z) z^{n_\nu}) = \sum_{\lambda=0}^{n_\nu - 1} z^\lambda \frac{1}{\pi} \int_N \tilde{\phi}_{k\nu}(\zeta)\, \zeta^{n_\nu - \lambda - 1}\, d\xi d\eta - (\underset{\sim}{T}\tilde{\phi}_{k\nu}) z^{n_\nu}$$

$$= \text{analytic function} + \phi_{k\nu}(z) z^{n_\nu}.$$

Likewise,

$$-\underset{\sim}{T}(\tilde{p}_{k\nu}(\tfrac{\bar{z}}{z}) z^{n_\nu - 1}) = \sum_{\lambda=0}^{n_\nu - 1} z^\lambda \frac{1}{\pi} \int_N \tilde{p}_{k\nu}(\tfrac{\bar{\zeta}}{\zeta}) \zeta^{n_\nu - \lambda - 2}\, d\xi d\eta$$

$$-(\underset{\sim}{T}\, \tilde{p}_{k\nu}(\tfrac{\bar{z}}{z})\tfrac{1}{z}) z^{n_\nu}$$

$$= \text{analytic} + p_{k\nu}(\tfrac{\bar{z}}{z}) \frac{\bar{z}}{z}\, z^{n_\nu}.$$

As $\tilde{\phi}_{k\nu} \in L_p(N)$, $p>2$, we have $\phi_{k\nu} \in C^\beta(N)$ for $0<\beta = \frac{p-2}{p} < \alpha$, and $\deg p_{k\nu} \le \min\{k-\nu-1, n_\nu - 1\}$. From this we have that

$$\phi_k(z) = \hat{\phi}_k(z) + \sum_{\nu=0}^{k-1} [\phi_{k\nu}(z) + p_{k\nu}(\tfrac{\bar{z}}{z})\tfrac{\bar{z}}{z}] z^{n_\nu}, \qquad (19)$$

where $\hat{\phi}_k$ is analytic in N and $\hat{\phi}_k(0) = 0$. In the case $\hat{\phi}_k(z) \not\equiv 0$ then there exist positive integers n_k such that $\hat{\phi}_k(z) \equiv z^{n_k} \phi_{kk}(z)$, where ϕ_{kk} is analytic in N and $\phi_{kk}(0) \neq 0$. When $\hat{\phi}_k(z) \equiv 0$, then we define $\phi_{kk}(z) \equiv 1$ and represent

$$\phi_k(z) = \phi_{kk}(z) z^{n_o} + \sum_{\nu=1}^{k-1} [\phi_{k\nu}(z) + p_{k\nu}(\tfrac{\bar{z}}{z})\tfrac{\bar{z}}{z}] z^{n_\nu} + (\phi_{ko}(z) + p_{ko}(z) - 1) z^{n_o}.$$

Hence, this establishes (15) for $\mu=k$, which means it holds for all $\nu \leq r-1$. Since one could make similar combinations for arbitrary $\nu (0 \leq \nu \leq k-1)$, it is clear that the n_k are not uniquely defined by the above procedure.

Lemma: In a neighborhood of the origin $N(0)$, $\phi_\mu(z)$ may be represented as

$$\phi_\mu(z) = \alpha_\mu z^{n_\mu} \phi_{\mu\mu}(z) + \sum_{\nu=0}^{\mu-1} \alpha_\nu \{\phi_{\mu\nu}(z) + p_{\mu\nu}(\tfrac{\bar{z}}{z}) \tfrac{\bar{z}}{z}\} z^{n_\nu}, \qquad (15')$$

where $\alpha_\mu \in \{0,1\}, n_\nu \in \mathbb{N}, \phi_{\mu\nu} \in C^\alpha(N) \cap D_{1,p}(N)$ $(0<\alpha<1, 2<p)$, $\mu=0,1,\ldots,r-1$. Moreover, $\phi_{\mu\mu}$ is analytic in N, $\phi_{\mu\mu}(0) \neq 0$, and $p_{\mu\nu}$ is a polynomial and $\deg p_{\mu\nu} \leq \min\{\mu-\nu-1, n_\nu-1\}$.

Proof: As in the previous proof by induction we arrive at an equation of the type (19). In the instance that $\hat{\phi}_k \not\equiv 0$ we choose $\alpha_k=1$. On the other hand, for $\hat{\phi}_k \equiv 0$ we take $\alpha_k=0$, $\phi_{kk}=1$, and $n_k := n_0$. This then yields (15').

From this representation of a solution ϕ of $D^*\phi = 0$ in the neighborhood of a zero, we obtain the local behavior of a solution of the first-order system (1), namely in the neighborhood of a zero z_0 of w, we have

$$w(z) = TS\hat{Z}, \qquad (20)$$

where T is the lower triangular matrix (9), $S(z)$ is the matrix with components $s_{\mu\nu}$ $(0 \leq \mu, \nu \leq r-1)$ defined by $s_{\mu\nu}=0$ for $\mu<\nu$, $s_{\mu\mu}=\phi_{\mu\mu}$,

$$s_{\mu\nu}(z) := \phi_{\mu\nu}(z) + p_{\mu\nu}\left(\frac{\bar{z}-\bar{z}_0}{z-z_0}\right) \frac{\bar{z}-\bar{z}_0}{z-z_0} \quad (\nu<\mu), \qquad (21)$$

and $\hat{Z}$ is the colum vector

$$\hat{Z} := (\alpha_0 (z-z_0)^{n_0}, \ldots, \alpha_{r-1}(z-z_0)^{n_{r-1}})', \qquad (22)$$

The previous lemma suggests a normalized representation. Let us suppose that $\alpha_\mu=1$ and $n_\mu \geq n_{\nu_o}$ for at least one $\nu_o (0 \leq \nu_o \leq \mu-1)$ where $\alpha_{\nu_o}=1$. In this case (15') may be put into a "reduced" form

$$\phi_\mu(z) = 0\cdot z^{n_\mu}\cdot 1 + \sum_{\nu=0}^{\mu-1} [\hat{\phi}_{\mu\nu}(z)+p_{\mu\nu}(\frac{\bar{z}}{z})\frac{\bar{z}}{z}]\alpha_\nu z^{n_\nu},$$

where

$$\hat{\phi}_{\mu\nu}:=\phi_{\mu\nu} \quad (\nu\neq\nu_o), \quad \hat{\phi}_{\mu\nu_o} := \phi_{\mu\nu_o} + \phi_{\mu\mu} z^{n_\mu - n_{\nu_o}}.$$

In this manner we may ensure that $n_\mu < n_\nu$, whenever $\alpha_\mu=\alpha_\nu=1$ for $0 \leq \nu < \mu \leq r-1$.

<u>Lemma</u>: <u>If the indices in the representation</u> (15') <u>for the</u> μ<u>th component</u> ϕ_μ <u>of the hypercomplex function</u> ϕ <u>obey the restriction</u>, $n_\mu < n_\nu$, <u>whenever</u> $\alpha_\mu=\alpha_\nu=1$ for $0 \leq \nu < \mu \leq r-1$, <u>then the</u> $\{\alpha_\mu, n_\mu\}$ $\mu=0,1,\ldots,r-1$ <u>are uniquely given</u>.

<u>Proof</u> Let us introduce the matrix components of S as

$$s_{\mu\nu} := \hat{\phi}_{\mu\nu} + p_{\mu\nu}\frac{\bar{z}}{z} \qquad (0 \leq \nu \leq \mu-1),$$

$$s_{\mu\mu} = \hat{\phi}_{\mu\mu} \quad (0 \leq \mu \leq r-1),$$

so that (15') may be written as

$$\phi_\mu = \sum_{\sigma=0}^{\mu} \alpha_\sigma s_{\mu\sigma} z^{n_\sigma}, \quad (0 \leq \mu \leq r-1).$$

Furthermore, let us suppose that a normalized representation exists having the form

$$\phi_\mu = \sum_{\sigma=0}^{\mu} \beta_\sigma t_{\mu\sigma} z^{m_\sigma}, \quad (0 \leq \mu \leq r-1),$$

where β_σ and m_σ fulfill the condition $m_\mu < m_\sigma$ whenever $\beta_\mu=\beta_\sigma=1$ for $0 \leq \sigma < \mu \leq r-1$. The $t_{\mu\sigma}$, furthermore, are bounded functions in $N(0)$, and the $t_{\mu\mu}(z)$ are analytic such that $t_{\mu\mu}(0)\neq 0$.

We consider next several cases which arise. If $\mu=0$ then we have $\alpha_o s_{oo} z^{n_o}=\beta_o t_{oo} z^{m_o}$ with analytic $s_{oo}(z)$ and $t_{oo}(z)$. This implies $\alpha_o=\beta_o$ and $n_o=m_o$ if $\alpha_o=\beta_o=1$. If $\alpha_o=\beta_o=0$ we proceed to $\mu=1$ and consider $\alpha_1 s_{11} z^{n_1} = \beta_1 t_{11} z^{m_1}$. Next we consider the general case and make the induction assumption that $n_\nu=m_\nu$ when $\alpha_\nu=\beta_\nu=1$ for $\nu<\mu$. The representation of ϕ_μ implies that

$$\alpha_\mu s_{\mu\mu} z^{n_\mu} - \beta_\mu t_{\mu\mu} z^{m_\mu} = \sum_{\nu=0}^{\mu-1} \{\beta_\nu t_{\mu\nu} z^{m_\nu} - \alpha_\nu s_{\mu\nu} z^{n_\nu}\}. \qquad (23)$$

Suppose now that $n_\mu=m_\mu$. This implies that

$$\alpha_\mu s_{\mu\mu} - \beta_\mu t_{\mu\mu} = \sum_{\nu=0}^{\mu-1} \{\beta_\nu t_{\mu\nu} z^{m_\nu - m_\mu} - \alpha_\nu s_{\mu\nu} z^{n_\nu - n_\mu}\},$$

which by our index restriction leads to

$$\lim_{z\to 0} (\alpha_\mu s_{\mu\mu} - \beta_\mu t_{\mu\mu}) = \alpha_\mu s_{\mu\mu}(0) - \beta_\mu t_{\mu\mu}(0) = 0.$$

Since $s_{\mu\mu}(0)$, and $t_{\mu\mu}(0)$ are not zero we must have $\alpha_\mu=\beta_\mu$. If, on the other hand, $n_\mu\neq m_\mu$, say $n_\mu<m_\mu$ then we proceed by induction to show $\alpha_\mu=\beta_\mu=0$. Using the induction hypothesis on (23) yields

$$\alpha_\mu s_{\mu\mu} z^{n_\mu - m_\mu} - \beta_\mu t_{\mu\mu} = \sum_{\nu=0}^{\mu-1} \{\beta_\nu t_{\mu\nu} - \alpha_\nu s_{\mu\nu}\} z^{m_\nu - m_\mu},$$

and hence

$$\lim_{z\to 0} \alpha_\mu s_{\mu\mu} z^{n_\mu - m_\mu} = \beta_\mu t_{\mu\mu}(0),$$

from which we conclude $\alpha_\mu=\beta_\mu=0$. Hence $\alpha_\mu=\beta_\mu$ in both cases and $\alpha_\mu=\beta_\mu=1$ is only possible if $n_\mu=m_\mu$.

Remark: In general $s_{\mu\nu}\neq t_{\mu\nu}$ since the vector $Z:=[\alpha_o z^{n_o},\dots,\alpha_{r-1} z^{n_{r-1}}]'$ may be a solution of the homogeneous algebraic system $(s_{\mu\nu}-t_{\mu\nu})\, Z(z)=0$ for each $z\in N(0)$.

If $\phi(z)$ solves $D^*\phi = 0$ and z_o be a zero of ϕ then in the neighborhood of z_o

$$\phi = SZ\ , \ Z:= (\alpha_o(z-z_o)^{n_o},\dots,\alpha_{r-1}(z-z_o)^{n_{r-1}})', \qquad (24)$$

where $S:=(s_{\mu\nu})$ is a triangular matrix of the type constructed above. Because (α_ν, n_ν) are uniquely determined

$$\min\{n_\nu : 0\le\nu\le r-1,\ \alpha_\nu=1\} \in \mathbb{N}$$

is a characteristic number which we shall call the <u>order of the</u> zero of ϕ at z_o. From our representation (11) this is also seen to be the order of $w=T\phi$.

Let us now consider the question of poles for hyperanalytic functions. It is conceivable that the complex part does not have a singularity but one of the nilpotent components does. For this reason we must require that in order for a point z_o to qualify as a pole of $w(z)$ the reciprocal function must have a zero. Following a modified form of the scheme used to investigate the local behavior around zeros, it is possible to show that the $\mu^{\underline{th}}$ component may be represented in the form

$$\phi_\mu(z) = \alpha_\mu z^{-n_\mu}\phi_{\mu\mu}(z) + \sum_{\nu=0}^{\mu-1}\{\phi_{\mu\nu}(z) + p_{\mu\nu}(\tfrac{\overline{z}}{z})\tfrac{\overline{z}}{z}\}\alpha_\nu z^{-n_\nu}, \tag{25}$$

$\mu=0,1.,,,,r-1$, where as before $\alpha_\mu\in\{0,1\}$ and $p_{\mu\nu}$ is a polynomial. To this end let us suppose this has been verified for all ν, with $0\le\nu\le\mu-1$, then

$$\frac{\partial\phi_\mu}{\partial\overline{z}} = -\sum_{k=0}^{\mu-1} q_{\mu k}\{\sum_{\nu=0}^{k}[\frac{\partial}{\partial z}(\phi_{k\nu})z^{-n_\nu} - n_\nu\phi_{k\nu}z^{-n_\nu-1}]$$

$$+\sum_{\nu=0}^{k-1}\hat{p}_{k\nu}(\tfrac{\overline{z}}{z})\tfrac{\overline{z}}{z}\,z^{-n_\nu-1}\}.$$

Following our earlier reasoning this leads to an expression of the form

$$\frac{\partial\phi_\mu}{\partial\overline{z}} = -\sum_{\nu=0}^{\mu-1}[\tilde{\phi}_{\mu\nu}(z)z^{-n_\nu} + \tilde{p}_{\mu\nu}(\tfrac{\overline{z}}{z})z^{-n_\nu-1}],$$

where $\tilde{\phi}_{\mu\nu}$ is Hölder continuous and $\tilde{p}_{\mu\nu}$ is a different polynomial. The function $\phi_\mu(z)$ may then be represented as a sum of functions $\psi_\nu(z)z^{-n_\nu}$, where the $\psi_\nu(z)$ are solutions of the equations

$$\frac{\partial\psi_\nu}{\partial\bar{z}} = \tilde{\phi}_{\mu\nu}(z) + \tilde{p}_{\mu\nu}(\frac{\bar{z}}{z})\frac{1}{z};$$

more precisely,

$$\phi_\mu(z) = \hat{\phi}_\mu(z) + \sum_{\nu=0}^{\mu-1} [\phi_{\mu\nu}(z) + p_{\mu\nu}(\frac{\bar{z}}{z})\frac{\bar{z}}{z}]z^{-n_\nu},$$

$\hat{\phi}_\mu(z) = \phi_{\mu\mu}(z)z^{-n_o}$ where $\phi_{\mu\mu}(z)$ is analytic and $\phi_{\mu\mu}(0)\neq 0$. The generalization to the form (25) then follows as before. It can be shown, furthermore, that if we assume that $n_\mu \leq n_{\nu_o}$ for at least one ν_o then following the earlier scheme we may normalize our exponents such that $n_\nu < n_\mu$ whenever $\alpha_\mu = \alpha_\nu = 1$ for all μ,ν with $0 \leq \nu < \mu \leq r-1$. Likewise, under these restrictions the $\{\alpha_\mu, n_\mu\}$ are unique. The proof is analogous to the case of zeros. As before it is also possible to choose an <u>order of the pole</u>. In strict analogy to the case of a zero we choose the order to be the characteristic number given by

$$\max\{n_\nu : 0 \leq \nu \leq r-1,\ \alpha_\nu = 1\} \in \mathbb{N}.$$

This definition chooses the order of the most singular polar component.

There is something lacking in the above definitions of order in that it is the asymptotic behavior around the zeros and poles of the complex component of a generalized hyperanalytic function which classifies the index of a Riemann boundary value problem. Hence, with this in mind we introduce as the <u>degree</u> of a zero or a pole (of a generalized hyperanalytic function) the order of the zero or pole of its complex part. Since $n_\mu > n_\nu$ for poles when $\mu > \nu$ the function

$$w(z) := [t(z) - t(z_o)]^{n_o} w(z)$$

may be singular if w(z) has a pole at z_o. Likewise, if w(z) has a zero at z_o the function

$$v(z) := [t(z) - t(z_o)]^{-n_o} w(z)$$

may also be singular at z_o.

IV. Solvability of the Problems ($\mathring{A}$) and (A).

In order to investigate the solvability of the boundary value problems it is useful to develop a reflection principle for the hypercomplex differential equation. We do this for domains $G\subset\{|z|<1\}$ having an arc for a portion of the boundary; more precisely, let $\gamma\subset\partial G\cap\{|z|=1\}$ be such an arc segment. Let w(z) be a solution of the special Riemann Hilbert problem

$$Dw + Aw + B\overline{w} = 0 \text{ in } G \qquad (1)$$
$$\mathrm{Re}\, w = 0 \quad \text{on} \quad \gamma.$$

Then it may be shown that it is possible to reflect w(z) across the segment γ into the reflection G_* of G. The reflection is performed by the identification

$$w_*(z) := \begin{cases} w(z) & , \quad z\in G \\ \overline{w(\frac{1}{\overline{z}})} & , \quad z\in G_* . \end{cases} \qquad (2)$$

Moreover, $w_*(z)$ satisfies the differential equation

$$D_*w_* + A_*w_* + B_*\overline{w}_* = 0 \qquad (3)$$

in the domain $G\cup G_*\cup\gamma$; on γ $w_*^+ = w_*^-$, and

$$A_*(z) := \begin{cases} A(z) & , \ z\in G \\ \overline{\frac{1}{z^2} A(\frac{1}{\overline{z}})} , & z\in G_* \end{cases} , \quad B_*(z) := \begin{cases} B(z) , & z\in G \\ \overline{\frac{1}{z^2} B(\frac{1}{\overline{z}})} , & z\in G_* \end{cases} , \qquad (4)$$

$$D_* := \frac{\partial}{\partial\overline{z}} + Q_*\frac{\partial}{\partial z} , \quad Q_*(z) := \begin{cases} Q(z) , & z\in G, \\ \frac{z^2}{\overline{z}^2}\, \overline{Q(\frac{1}{\overline{z}})} , & z\in G^* . \end{cases} \qquad (5)$$

That this is the case is seen by computing for $z\in G^*$

$$\frac{\partial}{\partial \bar z} w_*(z) + \Omega_*(z) \frac{\partial}{\partial z} w_*(z) = -\overline{\frac{\partial}{\partial z} w(\frac{1}{\bar z})} - \frac{z^2}{\bar z^2} \overline{\Omega(\frac{1}{\bar z})}\, \overline{\frac{\partial}{\partial \bar z} w(\frac{1}{\bar z})}$$

$$= \frac{1}{\bar z^2} [\overline{\frac{\partial}{\partial (1/z)} w(\frac{1}{\bar z})} + \overline{\Omega(\frac{1}{\bar z})}\, \overline{\frac{\partial}{\partial (1/\bar z)} w(\frac{1}{\bar z})}]$$

$$= -\frac{1}{z^2} [\overline{A(\frac{1}{\bar z})}\,\overline{w(\frac{1}{\bar z})} \qquad \overline{(\frac{1}{\bar z})}\; w(\frac{1}{\bar z})] = -[A_*(z) w_*(z) + B_*(z) \overline{w_*(z)}].$$

Furthermore, for $z \in \gamma$ we have $w_*^-(z) := -\overline{w(\frac{1}{\bar z})} = -\overline{w(z)}$, but because $\mathrm{Re}\, w = 0$, $w(z) = -\overline{w(z)}$ on γ.

We are now able to prove a variation on a decomposition theorem of the form found in Vekua (Th. 4.4 page 243).

<u>Theorem</u> : <u>Let</u> $\Gamma \in C^{1,\mu}$, $0<\mu\le 1$, <u>and</u> w <u>be</u> <u>a</u> <u>nontrivial</u> <u>solution</u> <u>of</u> ($\mathring{A}$). <u>Then</u> $w = P(t(z))\hat w$, <u>where</u> P <u>is</u> <u>a</u> <u>polynomial</u> <u>whose</u> <u>roots</u> (<u>as</u> <u>a</u> <u>function</u> <u>of</u> z) <u>all</u> <u>belong</u> <u>to</u> $G \cup \Gamma$; $\hat w \in C(G\cup\Gamma)$ <u>and</u> <u>the</u> <u>complex</u> <u>part</u> <u>of</u> $\hat w$, <u>namely</u> $\hat w_o$, <u>does</u> <u>not</u> <u>vanish</u> <u>in</u> G.

<u>Proof</u>: We first map G conformally onto G' where G' is a domain bounded by circles Γ' and such that the origin is an interior point of G'. Now if w satisfies the boundary value problem

$$Dw + Aw + B\bar w = 0 \quad \text{in } G,$$

$$\mathrm{Re}\lambda w = 0 \quad \text{on } \Gamma,$$

then the composed function $W(\zeta) := w(\phi(\zeta))$, where $\phi: G' \to G$, satisfies

$$\tilde D W + \tilde{\tilde A} W + \tilde{\tilde B} \bar W = 0 \quad \text{in } G',$$

$$\mathrm{Re}\, \lambda_1 W = 0 \quad \text{on } \Gamma',$$

where $\tilde D := \frac{\partial}{\partial \bar\zeta} + \frac{\partial}{\partial \zeta}$, $\tilde{\tilde A}(\zeta) := \overline{\phi'} A(\phi(\zeta))$, etc., $\tilde\Omega := \overline{\phi'}\Omega(\phi)(\phi')^{-1}$.
This follows directly from

$$\tilde D W := (\frac{\partial}{\partial \bar\zeta} + \frac{\partial}{\partial \zeta})\, W = \overline{\phi'}\, [\frac{\partial}{\partial \bar z} + \Omega \frac{\partial}{\partial z}] W$$

$$= -\overline{\phi'}\,AW - \overline{\phi'}\,B\overline{W}\ .$$

Letting $\tilde{t}(\zeta)$ be a generating solution associated with the operator $\tilde{D}$ we introduce

$$V(\zeta) := W(\zeta)\ \exp(-ip(\zeta))\ \prod_{k=1}^{m}[\tilde{t}(\zeta)-\tilde{t}(\zeta_k)]^{n_k}\exp(\underset{\sim}{T}\tilde{\tilde{A}})\ ,$$

and choosing $a_1=\ldots a_{|n|}=0$ (see (II. 14)) we may normalize $\tilde{t}(\zeta)$ such that the boundary data is given by

$$\operatorname{Re}(\overline{\tilde{t}(\zeta)^n}\ \exp(-i\pi\alpha)\ V)\ = 0 \quad \text{on}\ \ \Gamma',$$

(see (II. 37)) or as $|t|$ is hyperreal

$$\operatorname{Re}(\tilde{t}(\zeta)^{-n}\ \exp(-i\pi\alpha)\ V) = 0 \quad \text{on}\ \Gamma'.$$

The function $\omega(\zeta):=\tilde{t}(\zeta)^{-n}\ V(\zeta)$ is then seen to be a solution of the normalized differential equation

$$\tilde{D}\omega + \tilde{B}\ \overline{\omega}\ = 0,$$

where $\tilde{B}$ is in G' given by (I.38). Except perhaps in a small neighborhood of the origin this function is of class A_p, $p>2$. Furthermore, the solution is continuous on Γ' and as $\alpha(z)$ is constant on each circle of Γ, it is clear from the reflection principle, that $\tilde{t}(\zeta)^{-n}\ V(\zeta)$ is continuously continuable outside G'†. From the Similarity Principle developed in Section III it is clear that the continued function $\omega_*(\zeta)$ has only a finite number of zeros in $(G'\cup\Gamma')-\mathcal{N}(0)$. Hence, $V(\zeta)$ has only a finite number of zeros in $G'\cup\Gamma'$. We know from III that it is possible to define the degrees ν_k of the zeros $\tilde{\zeta}_k$ of the function V, and

---

† Q^* is not continuous across $\dot{G}$; however, as the first component of ω is analytic since Q^* is nilpotent the zeros of ω are isolated and finite in number. The continued function, moreover, can be shown to be Hölder continuous as $Q^*\in L_p$, $p>2$, in the extended region of definition.

therefore the function defined by

$$\hat{V}(\zeta) := \prod_{k=1}^{N} [\tilde{t}(\zeta)-\tilde{t}(\hat{\zeta}_k)]^{-\nu_k} V(\zeta), \qquad (6)$$

has the property that its complex component $\hat{V}_o \neq 0$ in $G' \cup \Gamma'$. This implies that

$$W(\zeta) = p_1(\zeta)\hat{W}(\zeta),$$

where $p_1(\zeta) := P_1(\tilde{t}(\zeta))$ is a polynomial in $\tilde{t}$, and the complex part of $\hat{W}$, $\hat{W}_o(\zeta) \neq 0$ in $G' \cup \Gamma'$. Transforming back to the z-variables we get the desired result, namely

$$w(z) = p_1(\phi^{-1}(z))\hat{W}(\phi^{-1}(z))$$

or

$$w(z) = \prod_{k=1}^{N} [t(z) - t(z_k)]^{\nu_k} \hat{w}(z) \qquad (7)$$

where $\hat{w}_o(z) \neq 0$ in $G \cup \Gamma$.

Let points $\{a_1, \ldots, a_\ell\}$ be the roots of the function $P(t(z))=0$, which lie inside G. Furthermore, let the points $\{a_{j1}, \ldots, a_{j\ell_j}\}$ be the roots of $P(t(z))=0$ belonging to the boundary component Γ_j $(j=0,1,\ldots,m)$. The indices ν_i, ν_{ji} are the multiplicities of the zeros of $P(t)$ associated with $t=t(a_i)$ and $t=t(a_{ji})$ respectively. Let us now denote the hypercomplex arguments of the functions λ, P, and $\hat{w}$ by

$$\theta := \arg \lambda, \quad \phi := \arg P, \quad \hat{\theta} := \arg \hat{w}, \qquad (8)$$

and the complex part of these arguments by θ_o, ϕ_o, and $\hat{\theta}_o$ respectively. Since the index of the Riemann-Hilbert problem depends only on the variation of the complex part of the data, and since for Problem (A) of ¢II. 1) we have $e^{i(\phi_o+\hat{\theta}_o-\theta_o)}=i(-1)^k$, k an integer, we obtain on Γ_j the relation

$$2\pi n_j = \int_{\Gamma_j} d\arg\lambda = \int_{\Gamma_j} d\arg\lambda_o = \int_{\Gamma_j} d\arg P_o + \int_{\Gamma_j} d\arg \hat{w}_o, \qquad (9)$$

$(j=0,1,\ldots,m)$, where P_o = complex part of P. On the outer boundary Γ_o it becomes

$$\int_{\Gamma_o} d\,\arg P_o = \pi \sum_{i=0}^{\ell_o} \nu_{oi} + 2\pi \sum_{i=1}^{\ell} \nu_i + 2\pi \sum_{j=1}^{m} \sum_{i=1}^{\ell_j} \nu_{ji}. \tag{10}$$

Introducing the indices $N_G := \sum_{i=1}^{\ell} \nu_i$, and $N_{\Gamma_j} := \sum_{i=1}^{\ell_j} \nu_{ji}$ we obtain for interior boundary curves Γ_j that

$$\int_{\Gamma_j} d\,\arg P_o = -\pi N_{\Gamma_j} \qquad (j=1,\ldots,m) \tag{11}$$

and for Γ_o the equation (10) may be rewritten as

$$\int_{\Gamma_o} d\,\arg P_o = \pi N_{\Gamma_o} + 2\pi [N_{\Gamma_1} + \ldots + N_{\Gamma_m}] + 2\pi\, N_G. \tag{12}$$

If the index of Problem (A) is n, that is if $\frac{1}{2\pi}\int_{\Gamma} d\,\arg\lambda = n$, then by summing (9) over all j and using (11) (12) we obtain

$$2n = N_{\Gamma_o} + N_{\Gamma_1} + \ldots + N_{\Gamma_m} + 2N_G + 2\hat{n}, \tag{13}$$

where $\hat{n} := \sum_{j=0}^{m} \hat{n}_j$, and $2\pi\hat{n}_j := \int_{\Gamma_j} d\,\arg\,\hat{w}$. (14)

Since from the Theorem the complex part of $\hat{w}$ does not vanish in $\hat{G} := G+\Gamma$, we have

$$\hat{n} := \frac{1}{2\pi}\int_{\Gamma} d\,\arg\,\hat{w} = \frac{1}{2\pi}\int_{\Gamma} d\,\arg\,\hat{w}_o = 0,$$

from which we obtain the analogues of the index relations holding in the generalized analytic case, namely

$$2n = 2N_G + N_\Gamma, \text{ where } N_\Gamma := N_{\Gamma_o} + N_{\Gamma_1} + \ldots + N_{\Gamma_m}. \tag{15}$$

Remarks: If the index n is negative then it was shown in Begehr-Gilbert [2] that

$$\hat{w}(z) := \prod_{k=0}^{n} [t(z)-t(z_k)]^{-1} w(t), \quad z_k \in G \tag{16}$$

satisfies the differential equation

$$D\hat{w} = \bar{A}\,\hat{w} + B^{*}\,\overline{\hat{w}} = 0, \tag{17}$$

where $B^{*} := \prod_{k=1}^{n} \left[\frac{\overline{t(z)} - \overline{t(z_k)}}{t(z) - t(z_k)}\right] \cdot B(z)$, and a Riemann-Hilbert boundary condition of index zero. It has been shown [2] that for (17) there exists exactly $2|n|+1$ linearly independent, continuous solutions over A_r with non-vanishing complex parts. Moreover, since the nilpotent components of $\hat{w}$ satisfy the equations

$$\frac{\partial w_k}{\partial \bar{z}} + \bar{A}_o \hat{w}_k + B_o^{*}\overline{\hat{w}}_k = -\sum_{\ell=0}^{k-1} [\bar{A}_{k-\ell}\hat{w}_\ell + B^{*}_{k-\ell}\overline{\hat{w}} + a_{k,k-\ell}\frac{\partial \hat{w}_\ell}{\partial z}]$$

$(k=1,\ldots,r-1)$, and as $\hat{w}_o(z) \in C^{\alpha}(\hat{G})$ it follows that the $\hat{w}_k$ are also $C^{\alpha}(\hat{G})$ solutions.

We consider next the case where the solutions are permitted to have a finite number of singular points in the domain and on its boundary. Moreover, we consider the case where in the vicinity of a singular point the solution $w(z)$ of $(\overset{o}{A})$ has a complex component $w_o(z)$ having the form

$$w_o(z) = 0(|z-z_o|^{-\nu}).$$

From Section III it is clear that for points z_o in the interior of G ν must be an integer. We shall show that this is also the case when z_o is on the boundary of G. We assume $z_o \in \Gamma'$, a segment of the boundary system, and without affecting the generality, as we can accomplish this with a conformal mapping, it may be assumed that $z_o=0$ and Γ' is the segment $[-\rho,\rho]$ on the real axis. Using the representation (III.11) for w a solution of $(\overset{o}{A})$, namely

$$w(z) = \sum_{k=0}^{n-1} \phi_k(z)\tau^{(k)}, \text{ where } D^*\phi = 0,$$

it is easy to see that arg $\tau_o^{(o)}$ = arg λ_o on Γ', or what is equivalent that arg $\tau^{(o)}$ = arg λ. Consequently, the function ϕ is D^*-analytic in the semicircle $K_p := \{z : |z| < \rho' < \rho,\ \mathrm{Im} z > 0\}$ and continuous in $\bar{K}_\rho \setminus \{0\}$, and its complex part has the local behavior $\phi_o = 0(|z|^{-\nu})$. Following Vekua [18] (pg. 247) we see that ϕ must satisfy the boundary condition $\mathrm{Re}[\phi] = 0$ on $\Gamma' \cap \bar{K}_\rho$. It can be shown that in K_ρ $\phi(z)$ must have the form $f(t^*(z))$ where $t^*(z)$ is a D^*-generating function normalized to be real on $y=0$. This suggests we may continue ϕ into the lower semi-circle by the following scheme

$$\hat{\phi}(z) := \begin{cases} \phi(z) := f(t^*(z)), & \mathrm{Im} z \geq 0 \\ -\overline{\phi(\bar{z})} = -\bar{f}\ (\overline{t^*}(z)) = -\bar{f}(\overline{t^*(\bar{z})}), & \mathrm{Im}\ z < 0. \end{cases} \tag{18}$$

If we introduce $\hat{t}(z)$ as

$$\hat{t}(z) := \begin{cases} t^*(z), & \mathrm{Im}\ z \geq 0 \\ \overline{t^*(\bar{z})}, & \mathrm{Im}\ z < 0, \end{cases} \tag{19}$$

then

$$\hat{\phi}(z) := \begin{cases} f(\hat{t}(z)), & \mathrm{Im}\ z \geq 0 \\ -\bar{f}(\hat{t}(z)), & \mathrm{Im}\ z < 0. \end{cases} \tag{20}$$

In this way it is seen that $\phi(z)$ has a polar singularity of degree ν at z_o. Hence, we may conclude that every solution of $\overset{o}{(A)}$ having a finite number of singularities on $\hat{G}$ may be represented as

$$w(z) = R\ (t^*(z))\ \hat{w}(z) \tag{21}$$

where R is a rational function, and $\hat{w}_o(z)$ is both zero-free and continuous. From this it is clear that the following relation holds between the number of poles and zeros of the complex part, $w_o(z)$, of a solution to $(\overset{o}{A})$:

$$2N_G + N_\Gamma - 2P_G - P_\Gamma = 2n \tag{22}$$

If for $n<0$ the number of poles of $w_o(z)$ where $w(z) \in (\overset{o}{A})$ inside G and on Γ satisfy $2P_G + P_\Gamma < -2n$, then $w(z) \equiv 0$ in $\hat{G}$. It is also clear that for $n=0$ that $N_G = N_\Gamma = 0$ and hence every continuous nontrivial solution to $(\overset{o}{A})$ has the form (See also in this regard Begehr-Gilbert [2].)

$$w(z) = c_o \hat{w}(z), \tag{23}$$

where $\hat{w}(z)$ vanishes nowhere in $\hat{G}$. Furthermore, for $n>0$ the continuous, nontrivial solution has zeros in $\hat{G}$; the number N_G of zeros inside G and the number of zeros N_Γ on Γ are related by $2N_G + N = 2n$. It is apparent that on each Γ_j there may exist only an even number of zeros. From this we see that the degree of the polynomial in <u>Theorem</u> satisfies

$$n \leq \deg P \leq 2n \tag{24}$$

We culminate our discussion by observing that the difference between the number of solutions of the homogeneous problem $(\overset{o}{A})$ and its adjoint problem $(\overset{o}{A}{}')$ is equal to the difference between the indices of these problems respectively, that is

$$\ell - \ell' = n - n' = 2n + 1 - m. \tag{25}$$

The proof of this follows as in the complex case; see in this regard Vekua [18] pp. 250-251.

References

1. Bers, L.: Theory of Pseudo-Analytic Functions, Courant Institute, New York, 1953.

2. Begehr, H., and Gilbert, R. P.: Randwertaufgaben ganzzahliger Charakteristik für verallgemeinerte hyperanalytische Funktionen, Applicable Analysis, Vol. 6, 189-205, 1977.

3. Begehr, H., and Gilbert, R. P.: On Riemann boundary value problems, J. Diff. Equats. 32, (1979) 1-14.

4. Begehr, H., and Gilbert, R. P.: Piecewise continuous solutions of pseudoparabolic equations in two space dimensions, Proc. Royal Soc. Edinbrugh, 81A, (1978) 153-173.

5. Bojarski, B. B.: The theory of generalized analytic vectors (in Russian), Annales Polonici Mathematicis, 17 (1966), 281-320.

6. Buchanan, J. and Gilbert, R. P.: The Hilbert problem for hyperanalytic functions, Appl. Analysis, 11 (1981) 303-323.

7. Douglis, A.: A function-theoretic approach to elliptic systems of equations in two variables, Comm. Pure. Applied Math 6 (1953) 259-289.

8. Gilbert R. P.: Constructive Methods for Elliptic Equations, Lecture Notes in Math. No. 365, Springer, Berlin, 1974.

9. Gilbert, R. P. and Hile, G.: Generalized hypercomplex function theory, Trans. Am. Math. Soc., 195 (1974), 1-29.

10. Gilbert, R. P. and Hile, G.: Hypercomplex function theory in the sense of L. Bers, Math. Nachrichten, 72 (1976), 187-200.

11. Gilbert, R. P. and Hile, G.: Hilbert function modules with reproducing kernels, Nonlinear Analysis, 1 (1977), 135-150.

12. Gilbert, R. P. and Wendland, W.: Analytic, generalized hyperanalytic function theory and an application to elasticity, Proc. Royal Soc. Edinburgh, 73A (1975), 317-331.

13. Habetha, K.: On zeros of elliptic systems of first order in the plane, in "Function Theoretic Methods in Differential Equations," Pitman, London, 1976.

14. Hile, G.: Hypercomplex function theory applied to partial differential equations. Ph.D. Dissertation, Indiana University (1972).

15. Kühn, E.: Über die Funktionentheorie und das Ähnlichkeit sprinzip einer Klasse elliptischer Differentialgleichungssysteme in der Ebene, Dissertation, Dortmund, 1974.

16. Pascali, D. Vecteurs analytiques généralisés, Rev. Roum. Math. pura appl., 10 (1965), 779-808.

17. Tutschke, W.: Partielle Komplexe Differentialgleichungen VEB Berlin, 1977.

18. Vekua, I.N.: Generalized Analytic Functions, Pergamon, London, 1962.

19. Wendland, W.: Elliptic Systems in the Plane, Pitman, London, 1979.

Contemporary Mathematics
Volume 11, 1982

COUPLED VARIATIONAL INEQUALITIES FOR FLOW FROM A NON-SYMMETRIC DITCH

John C. Bruch, Jr.
and
James M. Sloss
University of California, Santa Barbara

1. INTRODUCTION

This paper presents a numerical approach to solving flow through a homogeneous media from a non-symmetric ditch suggested by work of J. Remar (see Remar et al. [4]). There are two free boundaries involved, one on each side of the ditch. Use is made of Baiocchi extensions to a fixed larger region. A variational inequality for the extended region could be derived; however, it would not be coercive. The difficulty arises with the fixed boundary.

The extended region is separated into three regions: a region III that involves the fixed boundary and causes the difficulty with the variational inequality formulation, but does not cause difficulty when treated as a fixed boundary problem, and two regions I and II involving the free boundaries, with simple enough extensions so that there is no difficulty with the corresponding variational inequality. Regions I and II must intersect region III in open sets. The original single problem is broken into three coupled problems. A Schwarz alternating procedure is set up that can be solved on each region. Under the assumption that there is an overlap region for all of the successive solutions, it is shown that the procedure converges and gives the solution and that the free boundaries are monotonic analytic curves. A uniqueness theorem for the original problem is also given. Finally, numerical results for a special case are given along with computer times and the number of iterations for the numerical scheme. These results show the efficiency and simplicity of this new approach.

0271-4132/82/0000-0141/$06.50

2. DIFFERENTIAL EQUATION FORMULATION

The physical problem investigated is seepage from a non-symmetric ditch or channel through a porous medium underlain by a drain at a finite depth (Fig. 1). The porous medium is assumed to be homogeneous and isotropic. Capillary and evaporation effects are neglected. We assume Darcy's law holds, namely in the flow region Ω, the piezometric head is

$$\phi(x,y) = y + \frac{1}{\gamma}p(x,y), \qquad \gamma = \text{constant},\ p = \text{pressure}$$

If the channel is described by $y = \alpha(x)$, we consider $\alpha(x)$ extended to the right and left of the channel in a smooth way as indicated in Fig. 1. Let R $R = \{(x,y)\ \ 0 < y < \alpha(x),\ x_{22} < x < x_{12}\}$ where we set $a_i = (x_{1i}, y_{1i})$, $b_i = (x_{2i}, y_{2i})$ $\quad i = 1,2,3,4$

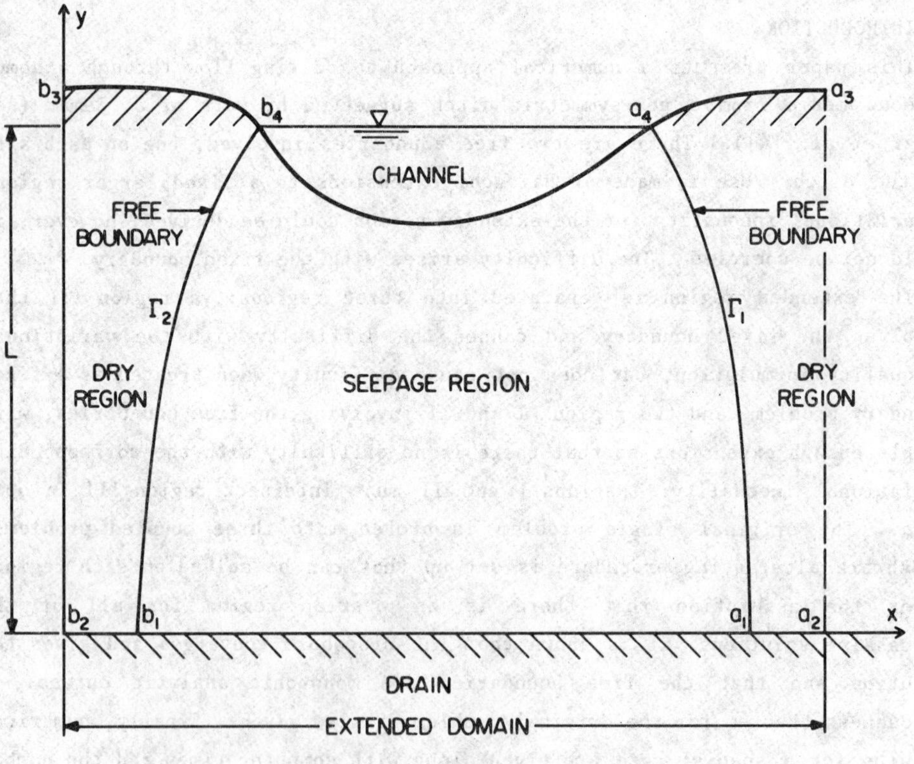

Figure 1. Flow domain and extended solution domain.

The model problem becomes with the water in the ditch at height L

<u>Problem 1</u>: Find $\{q, \Omega, \phi\}$ such that

(i) Ω is an open subset of R (Ω = seepage region)

(ii) $\Delta\phi = 0$ in Ω and if ψ is a conjugate of φ in Ω, $\phi, \psi \in C^0(\bar{\Omega})$

(iii) $\phi = y$, $\psi = 0$ on $\Gamma_1 \subset \partial\Omega$,

(iv) $\phi = y$, $\psi = q$ on $\Gamma_2 \subset \partial\Omega$, ($q$ = flowrate = constant)

(v) $\phi(x,0) = 0$

(vi) $\phi(x,\alpha(x)) = L$ = constant.

Remark: By the Hopf maximum principle $y - \phi < 0$ and $0 < \psi < q$ in Ω.

3. ω REFORMULATION

Given that $\{q,\Omega,\phi\}$ is a solution of Problem 1, we extend ϕ and ψ continuously to R by setting

$$\tilde\phi = \phi \;,\quad \tilde\psi = \psi \qquad \text{in } \Omega$$
$$\tilde\phi = y \;,\quad \tilde\psi = 0 \qquad \text{in } R - \Omega \text{ to the right of } \Gamma_1 \tag{3.1}$$
$$\tilde\phi = y \;,\quad \tilde\psi = q \qquad \text{in } R - \Omega \text{ to the left of } \Gamma_2$$

Note that

$$\tilde\phi_x - \tilde\psi_y = 0 \qquad \text{and} \qquad \tilde\phi_y + \tilde\psi_x = \chi_{R-\bar\Omega} \text{ on } R \tag{3.2}$$

in the sense of distributions where $\chi_{R-\bar\Omega}$ is the characteristic function on the set $R-\bar\Omega$. Let

$$\omega(x,y) = \int_{a_4}^{P} (y - \tilde\phi)dy - \tilde\psi dx \qquad P \in R$$

From (3.2) it follows that ω is independent of the path in R. Moreover $\omega > 0$ on Ω by the remark of § 2. For $(x,y) \in R^1 - \Omega$, $\omega = 0$, where $R^1 = \{(x,y) \in R \mid x > x_{14}\}$. For $(x,y) \in R^2 - \Omega$, $\omega = \omega(b_4) + q(x_{24} - x)$ as is seen by integrating from b_4 over to $x = x_{23}$ then down along x_{23} to (x_{23},y) and then horizontally back to (x,y) in $R^2 - \Omega$, where $R^2 = \{(x,y) \in R \mid x < x_{24}\}$. Set $A = \omega(b_4)$. By continuing to integrate horizontally, we conclude from the fact that $0 < \psi < q$ that

$$\omega > A + q(x_{24} - x) \text{ in } \Omega$$

Hence $\{q,\Omega,\phi\}$ is a solution of Problem 1 is equivalent to having $\{A,q,\Omega,\omega\}$ be a solution of:

Problem 2: Find $\{A,q,\Omega,\omega\}$ such that

(i) $\omega \geqq 0$ on R

(ii) $\Omega^1 = \{(x,y) \in R^1 \mid \omega(x,y) > 0\}$, $\Omega^2 = \{(x,y) \in R^2 \mid \omega(x,y) > A + q(x_{24} - x)\}$, $\Omega = \Omega^1 \cup \Omega^2 \cup \{(x,y) \mid 0 < y < \alpha(x),\ x_{24} \leq x \leq x_{14}\}$

(iii) $(\Delta\omega - 1)\omega = 0$ in R for which $x \geqq x_{24}$, $(\Delta\omega - 1)\cdot[\omega - A - q(x_{24} - x)] = 0$ in R for which $x \leqq x_{14}$, $\omega \in C^1(\bar R)$

(iv) $\omega_y(x,0) = 0$, $\omega(x_{12},y) = 0$, $\omega(x_{22},y) = A + q(x_{24} - x_{22})$

(v) $\omega(x,\alpha(x)) = A + q(x_{24} - x)$ for $x_{23} \leqq x \leqq x_{24}$

$= 0$ for $x_{14} \leqq x \leqq x_{13}$

$\omega_y(x,\alpha(x)) = y - L$ for $x_{24} \leqq x \leqq x_{14}$

3. A COUPLED PROBLEM

In this section we shall assume that there is an open subset S^1 of Ω to the right of $x = x_{14}$, and an open subset S^2 of Ω to the left of $x = x_{24}$, where $\partial S^j = \gamma'_j \cup \gamma_j \cup \{(x,y) \mid y = 0, x_{j5} \leq x \leq x'_{j5}\}$, $j=1,2$, see Fig. 2.

Let
$$R^4 = \{(x,y) \in R \mid x_{24} \leq x \leq x_{14}, \ 0 < y < \alpha(x)\}$$
$$R^3 = R^4 \cup S^1 \cup S^2$$

Our assumption states that $S^1 \subset R^1$ and $S^2 \subset R^2$.

Consider the following coupled problem.

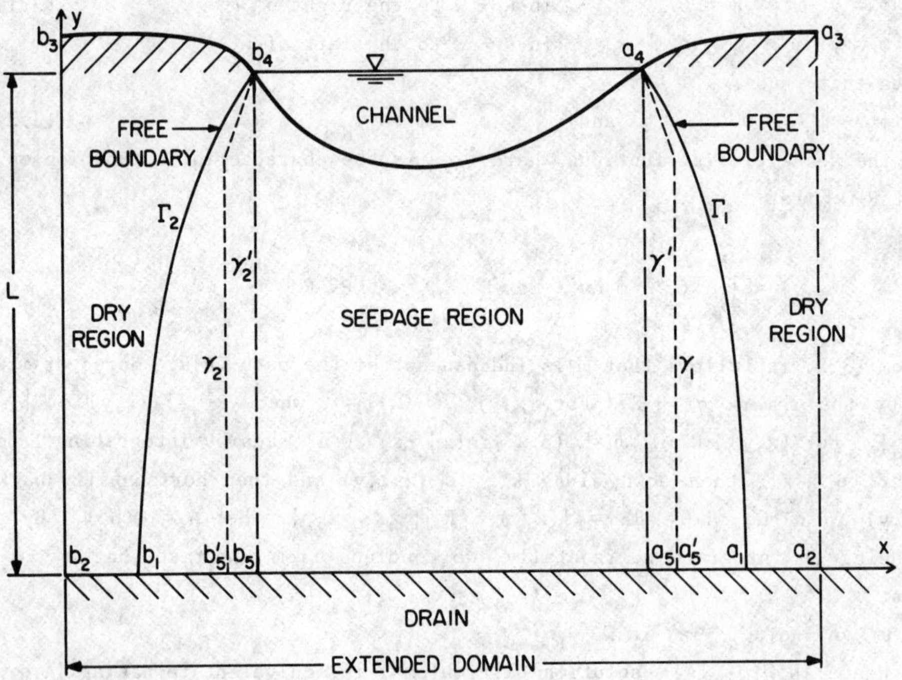

Figure 2. Coupled problem solution domains.

Problem 3: Find $\{A, q, \Omega, \omega^1, \omega^2, \omega^3\}$ such that

  (i) $\omega^j \geq 0$ in R^j, $\omega^j \in C^1(\bar{R}^j)$, $j = 1, 2, 3$, $\omega^2 \geq A + q(x_{24} - x)$

  (ii) $\Omega^1 = \{(x,y) \in R^1 \mid \omega^1 > 0\}$
  $\Omega^2 = \{(x,y) \in R^2 \mid \omega^2 > A + q(x_{24} - x)\}$
  $\Omega = \Omega^1 \cup \Omega^2 \cup R^4$

  (iii) $(\Delta\omega^1 - 1)\omega^1 = 0$ on R^1, $B^1[\omega^1] = F^1[x,y,\omega^3]$ on ∂R^1

  (iv) $(\Delta\omega^2 - 1)[\omega^2 - A - q(x_{24} - x)] = 0$ on R^2, $B^2[\omega^2] = F^2[x,y,\omega^3]$ on ∂R^2

  (v) $\Delta\omega^3 = 1$ on R^3, $B^3[\omega^3] = F^3[x,y,\omega^1,\omega^2]$ on ∂R^3

in which

$$
\begin{aligned}
B^1[\omega^1] &= \omega^1_y, & F^1[x,y,\omega^3] &= 0 \text{ on } a_5a_2 \\
&= \omega^1, & &= 0 \text{ on } a_2a_3 \\
&= \omega^1, & &= 0 \text{ on } a_3a_4 \\
&= \omega^1, & &= \omega^3 \text{ on } \gamma_1' \\
B^2[\omega^2] &= \omega^2_y, & F^2[x,y,\omega^3] &= 0 \text{ on } b_5b_2 \\
&= \omega^2, & &= A+q(x_{24}-x_{22}) \text{ on } b_2b_3 \\
&= \omega^2, & &= A+q(x_{24}-x) \text{ on } b_3b_4 \\
&= \omega^2, & &= \omega^3 \text{ on } \gamma_2' \\
B^3[\omega^3] &= \omega^3, & F^3[x,y,\omega^1,\omega^2] &= \omega^1 \text{ on } \gamma_1 \\
&= \omega^3_y, & &= y - L \text{ on } a_4b_4 \\
&= \omega^3, & &= \omega^2 \text{ on } \gamma_2 \\
&= \omega^3_y, & &= 0 \text{ on } b_5' \ a_5'
\end{aligned}
$$

Having a solution $\{A, q, \Omega, \omega^1, \omega^2, \omega^3\}$ of problem 3 is equivalent to having a solution of problem 2. To see this, on the overlap, e.g. $R^3 \cap R^1$, we have $\omega^3-\omega^1$ is harmonic with zero Dirichlet data on $\gamma_1 \cup \gamma_1'$ and hence $\omega^3-\omega^1 = 0$ on the overlap and consequently in $R^4 \cup R^1$. A similar conclusion holds for $R^4 \cup R^2$ and hence if

$\omega = \omega^j$ in R^j $\quad j = 1, 2, 3$

ω is defined on Ω.

Remark: Let $\bar{\omega} \equiv \omega_3-\omega_1$. By the Hopf maximum principle, the function $\bar{\omega}$, harmonic in $R^3 \cap R^1$, cannot take on its maximum or minimum where $\bar{\omega}_y=0$, hence $\bar{\omega}$ must take on its maximum and its minimum 0. A similar argument holds for $\omega_3 - \omega_2$.

4. UNIQUENESS

We are now in a position to prove that problem 1 has at most one solution.

Theorem (Uniqueness). Fix q, then problem 1 has at most one solution $\{q, \Omega, \phi\}$ for which $\partial\Omega \cap R^1$ and $\partial\Omega \cap R^2$ are smooth curves.

Proof: Assume $\{q, \Omega_1, \phi_1\}$ and $\{q, \Omega_2, \phi_2\}$ are two solutions. Let $\{A, q, \Omega_1, \omega_1\}$ and $\{A, q, \Omega_2, \omega_2\}$ be the two corresponding solutions of problem 2. Let

$$\bar{\omega} = \omega_1 - \omega_2 \quad \text{on } R$$

Then

$$\Delta\bar{\omega} = 0 \text{ on } \Omega_1 \cap \Omega_2,\ \Delta\bar{\omega} = \Delta\omega_1 = 1 \text{ on } \Omega_1 - \bar{\Omega}_2,\ \Delta\bar{\omega}_2 = -1 \text{ on } \Omega_2 - \bar{\Omega}_1$$

$$\bar{\omega}_y = 0 \text{ on } y = \alpha(x),\ \bar{\omega}_y = 0 \text{ on } y = 0,$$

$$\bar{\omega} = \omega_1 > 0 \text{ on } (\Omega_1 - \bar{\Omega}_2) \cap R^1,\ \bar{\omega} = -\omega_2 < 0 \text{ on } (\Omega_2 - \bar{\Omega}_1) \cap R^1$$

Let

$$\Gamma_1^1 = \{(x,y) \in \partial[(\Omega_1 - \bar{\Omega}_2) \cap R^1] \mid \omega_1 > 0,\ \omega_2 = 0\}$$

$$\Gamma_2^1 = \{(x,y) \in \partial[(\Omega_2 - \bar{\Omega}_1) \cap R^1] \mid \omega_2 > 0,\ \omega_1 = 0\}$$

$$\Gamma_3^1 = \{(x,y) \in \partial[(\Omega_1 \cup \Omega_2) \cap R^1] \mid \omega_1 = \omega_2 = 0\}$$

By assumption Γ_1^1, Γ_2^1, Γ_3^1 are collections of smooth arcs with $\Gamma_1^1 \subset \bar{\Omega}_1$, $\Gamma_2^1 \subset \bar{\Omega}_2$.

Note that $\Gamma_1^1 \subset \partial(\bar{\Omega}_1 \cap \bar{\Omega}_2) \cap \partial(\Omega_1 - \bar{\Omega}_2)$

$$\Gamma_2^1 \subset \partial(\bar{\Omega}_1 \cap \bar{\Omega}_2) \cap \partial(\Omega_2 - \bar{\Omega}_1)$$

It follows that $\bar{\omega}$ cannot take on a positive maximum on $\Gamma_1^1 \cup \Gamma_2^1 \cup \Gamma_3^1$. Indeed, if it did at p then $p \in \Gamma_1^1$ since on Γ_2^1 $\bar{\omega}<0$. Moreover

$$\Delta\bar{\omega} = 0 \text{ in } \Omega_1 \cap \Omega_2$$

from the Hopf maximum principle it would follow since $\bar{\omega} \in C^1(\bar{R})$ that

$$\frac{\partial\bar{\omega}}{\partial n}(p) = \frac{\partial(\omega_1 - \omega_2)}{\partial n} = \frac{\partial\omega_1}{\partial n} > 0$$

where n is an external normal to $\Omega_1 \cap \Omega_2$. However, since

$$\Delta\bar{\omega} = \Delta\omega_1 = 1 \text{ in } \Omega_1 - \bar{\Omega}_2$$

and since $\bar{\omega}=0$ on $\partial(\Omega_1-\bar{\Omega}_2)-(\Gamma_1^1 \cup \Gamma_2^1)$, it follows since $\omega_1 \in C^1(\bar{R})$ that

$$\max_{(\Omega_1 - \bar{\Omega}_2) \cap R^1} \bar{\omega}_1 = \max_{\partial[(\Omega_1 - \bar{\Omega}_2) \cap R^1]} \omega_1 = \max_{\Gamma_1^1} \omega_1 = \omega_1(p) = \bar{\omega}(p)$$

Thus since n is an inner normal to $(\Omega_1 - \bar{\Omega}_2)$, by the maximum principle since Γ_1^1 is smooth

$$\frac{\partial \omega_1}{\partial n}(p) < 0$$

which is a contradiction, and hence $\bar{\omega}$ cannot take on a positive maximum on $\Gamma_1^1 \cup \Gamma_2^1 \cup \Gamma_3^1$. In a similar way, we conclude that $\bar{\omega}$ cannot take on a negative minimum on $\Gamma_1^1 \cup \Gamma_2^1 \cup \Gamma_3^1$. A similar argument holds for the left side. Hence $\bar{\omega}$ cannot take on its maximum or minimum on $\partial(\Omega_1 \cap \Omega_2)$.

Since $\bar{\omega}_y = 0$ on $y = 0$, $\bar{\omega}$ cannot take on its maximum or minimum there. Hence

$$\max_{\overline{\Omega_1 \cup \Omega_2}} \bar{\omega} = \max_{y=\alpha(x)} \bar{\omega} = 0$$

and

$$\min_{\overline{\Omega_1 \cup \Omega_2}} \bar{\omega} = \min_{y=\alpha(x)} \bar{\omega} = 0$$

hence $\bar{\omega}(x,y) = 0$ in $\overline{\Omega_1 \cup \Omega_2}$ and the result follows.

5. VARIATIONAL INEQUALITIES FOR COUPLED PROBLEMS

Before we describe a sequence of coupled problems, it is convenient to introduce certain ideas. Introduce the following Hilbert spaces

$$M^1 = \{v^1 \in H^1(R^1) \mid v^1 = 0 \text{ on } a_2a_3,\ a_3a_4\}$$

$$M^2 = \{v^2 \in H^2(R^2) \mid v^2 = 0 \text{ on } b_2b_3,\ b_3b_4\}$$

and the following closed convex subsets:

$$K^1[g^1] = \{v^1 \in M^1 \mid v^1 \geqq 0,\ v^1 - g^1 = 0 \text{ on } \gamma_1'\},\ g^1 \in H^1(R^1),\ g^1 \geq 0 \text{ on } \partial R^1$$

$$K^2[\beta, g^2] = \{v^2 \in M^2 \mid v^2 \geqq \beta,\ v^2 - g^2 = 0 \text{ on } \gamma_2'\},\ g^2 \in H^1(R^1),\ g^2 \geq 0 \text{ on } \partial R^2,\ \Delta\beta = 0 \text{ on } R^2.$$

Lemma 5.1. If u^1 is defined on R^1 with $u^1 \in C^1(\bar{R}^1) \cap K^1[g^1]$ and if u^2 is defined on R^2 with $u^2 \in C^1(\bar{R}^2) \cap K^2[\beta, g^2]$ and if

(5.1) $(\Delta u^1 - 1)u^1 = 0$ on R^1, $B^1[u^1] = F^1[x,y,g^1]$ on ∂R^1

(5.2) $(\Delta u^2 - 1)(u^2 - \beta) = 0$ on R^2, $B^2[u^2] = F^2[x,y,g^2]$ on ∂R^2

then

(5.3) $a_1(u^1, v^1 - u^1) \geqq L_1(u^1, v^1 - u^1)$ for all $v^1 \in K^1[g^1]$

(5.4) $a_2(u^2, v^2 - u^2) \geqq L_2(u^2, v^2 - u^2)$ for all $v^2 \in K^2[\beta, g^2]$

with $L_j(u^j, v^j - u^j) = -\iint_{R^j} (v^j - u^j)\,dxdy$, $j = 1,2$

and $a_j(u^j, v^j - u^j) = \iint_{R^j} \nabla u^j \cdot \nabla(v^j - u^j)dxdy$

Proof (lemma)

Let

$$\Omega^1 = \{(x,y) \in R^1 \mid u^1 > 0\},\ \Omega^2 = \{(x,y) \in R^2 \mid u^2 > \beta\}$$

By Green's theorem

$$\iint_{R^1} [f\Delta g + \nabla f \cdot \nabla g]dxdy = \int_{\partial R^1} f\frac{\partial g}{\partial n}\, dS,\ n = \text{exterior normal}.$$

Note that on ∂R^1, $(v^1 - u^1)\frac{\partial u^1}{\partial n} = 0$. Since $\Delta w^1 = \chi_{\Omega^1}$ in the sense of distributions on R^1

$$a_1(u^1, v^1 - u^1) = -\iint_{\Omega^1} (v^1 - u^1)dxdy$$

However, since $v^1 \geqq 0$ and $u^1 = 0$ on $R^1 - \Omega^1$

$$\iint_{R^1-\Omega^1}(v^1 - u^1)dxdy \geqq 0$$

and the first inequality follows.

As for the second inequality, note that $(v^2 - u^2)\frac{\partial u^2}{\partial n} = 0$ on ∂R^2, hence since $\Delta u^2 = \chi_{\Omega^2}$ in the sense of distributions on R^2

$$a_2(u^2, v^2 - u^2) = -\iint_{\Omega^2} (v^2 - u^2)dxdy$$

However, since $v^2 \geqq \beta$ on R^2 and $u^2 = \beta$ on $R^2 - \Omega^2$

$$\iint_{R^2-\Omega^2}(v^2 - u^2)dxdy \geqq 0$$

and the second inequality holds.

We shall next state a lemma from Kinderlehrer and Stampacchia [3].

Lemma 5.2. Let $g^1 \in H^{2,s}(\Omega^1)$; $\beta, g^2 \in H^{2,s}(\Omega^2)$, $\beta \leqq g^2$ on b_2b_3, b_3b_4, for some s, $2 < s < \infty$. Then for the solutions (u^1,u^2) of (5.1), (5.2), u^j $H^{2,s}(\Omega^j) \cap C^{1,\lambda}(\bar{\Omega}^j)$, $j = 1,2$, $\lambda = 1-2/s$.

6. DESCRIPTION OF THE COUPLED ALTERNATING SCHEME

We wish now to describe a scheme for finding an approximate solution. Toward this end we give a sequence of coupled problems:

1. Let $A > 0$, $q > 0$ be fixed constants.

2. Guess ϕ_0^3 on $\bar{R}$ for which $y-\phi_0^3 \leqq 0, L \geqq \phi_0^3 \geqq 0$ on $\bar{R}$,

$$y-\phi^3 < 0 \text{ on } \gamma_1'+\gamma_2',\ \phi_{0,y}^3 \geq 0 \text{ on } \gamma_1' \text{ and } \gamma_2',\ \phi_0^3 \in C(R^1),\ \phi_0^3 \in H^{2,s}(R)$$

3. (a) Construct ψ_0^3 on $\bar{R}$, the harmonic conjugate of ϕ_0^3 with $\psi_0^3(a_4) = 0$.

   (b) Let $P \in \bar{R}$

$$\omega_0^3(P) = \int_{a_4}^{P} (y - \phi_0^3)dy - \psi_0^3 dx.$$

4. Solve the variational inequality (5.3), (5.4) with

$$g^1 = \omega_0^3,\ g^2 = \omega_0^3,\ \beta = A+q(x_{24}-x) \text{ for } (\omega_1^1, \Omega^1) \text{ on } R^1$$

and (ω_1^2, Ω_1^2) on R^2.

5. Set $\phi_1^1 = y-\omega_{1,y}^1$ on $R^1 \supset \Omega_1^1$, $\Omega_1^1 = \{(x,y) \in R^1 \mid \omega_1^1 > 0\}$

$$\phi_1^2 = y - \omega_{1,y}^2 \text{ on } R^2 \supset \Omega_1^2,\quad \Omega_1^2 = \{(x,y) \in R^2 \mid \omega_1^2 > \beta\}$$

Remark: Since $y-\phi_1^1 = y-\phi_0^3 < 0$ on γ_1' and ϕ_1^1 is continuous on R^1 it follows that there is a curve $\gamma_{1,1}$ joining a_4 to y=0 lying to the right of γ_1' with $\gamma_{1,1} \subset \Omega_1^1$ and $y-\phi_1^1 < 0$. Similarly there exists curve $\gamma_{2,1}$ joining b_4 to y = 0 with $\gamma_{2,1} \subset \Omega_1^2$. Assume the curves intersect y=0 at right angles and $\gamma_{1,1}$, $\gamma_{2,1}$ are composed of straight lines, one vertical meeting y = 0 and one diagonal meeting (x_{14}, y_{14}) (for $\gamma_{1,1}$) and meeting (x_{24}, y_{24}) (for $\gamma_{2,1}$).

6. Set $\phi_1^3 = \phi_1^1$ on $\gamma_{1,1}$

$\phi_1^3 = \phi_1^2$ on $\gamma_{2,1}$.

7. Let R_1^3 be the region bounded by $y = \alpha(x)$, $y = 0$, $\gamma_{1,1}$ and $\gamma_{2,1}$. Solve

$\Delta\phi_1^3 = 0$ in R_1^3, $\phi_1^3 = L$ on $y = \alpha(x)$, $\phi_1^3 = 0$ on $y = 0$
and boundary conditions of 6 on $\gamma_{1,1}$, $\gamma_{2,1}$.

Remark 1. $y - \phi_1^3 \leqq 0$ in R_1^3. This follows since not only does this hold for the harmonic function $y - \phi_1^3$ on $y = \alpha(x)$ and y=0, but also on $\gamma_{1,1}$ and $\gamma_{2,1}$. Indeed on Ω_1^1, $\omega_{1,y}^1 = y - \phi_1^1 = 0$ on $\partial\Omega_1^1 \cap R^1$ and on y=0, and since $y - \phi_1^1 = y - \phi_0^3 < 0$ on γ_1', by the Hopf maximum principle, $y - \phi_1^3 = y - \phi_1^1 \leqq 0$ on $\gamma_{1,1}$. Similarly for $\gamma_{2,1}$.

Remark 2. $y - \phi_1^3 < 0$ on γ_1' and γ_2'. This follows from above and the maximum principle and the fact that $y - \phi_1^3 = 0$ at some boundary point.

Remark 3. $L \geqq \phi_1^3 \geqq 0$ on R_1^3. The fact that $\phi_1^3 \geqq 0$ follows from remark 1. Since $\phi_1^1 = \phi_0^3 \leqq L$ on γ_1' and since $\phi_1^1 = y \leqq L$ on $\partial\Omega^1 \cap R^1$ and since $\phi_1^1 = 0$ on y=0 and since ϕ_1^1 is harmonic in Ω_1^1 it follows from the maximum principle that $\phi_1^3 = \phi_1^1 \leqq L$ on $\gamma_{1,1}$. Similarly $\phi_1^3 \leqq L$ on $\gamma_{2,1}$ and the result follows from the maximum principle applied to the function ϕ_1^3 harmonic on R_1^3.

Remark 4. $\omega_{1,y}^1 \leq 0$ on R^1 and $\omega_{1,y}^2 \leq 0$ on R^2. This follows from the maximum principle since $\omega_{1,y}^1 = 0$ on $\partial\Omega_1^1 \cap R^1$ and on y = 0 and $\omega_{1,y}^1 = \omega_{0,y}^3 \leq 0$ on γ_1' and $\omega_{1,y}^1$ is harmonic in Ω_1^1. Similarly for $\omega_{1,y}^2$.

Remark 5. $\omega^1_{1,x} \le 0$ on R^1, $\omega^2_{1,x} \ge -q$ on R^2. Note that $\omega^1_{1,x} = 0$ on $\partial\Omega^1_1 \cap R^1$. $\omega^1_{1,xy} = (\omega^1_{1,y})_x = 0$ on $y = 0$ and hence cannot take on its maximum on $y = 0$. On γ_1' $\omega^1_{1,xx} = 1 - \omega^1_{1,yy} = \phi^1_{1,y} = \phi^3_{0,y} \ge 0$ and hence $\omega^1_{1,x}$ can not take on its maximum ($\omega^1_{1,xx}$ interior normal) on γ_1'. Thus $\omega^1_{1,x} \le 0$ on R^1. Similarly, using the maximum principle it follows that $\omega^2_{1,x} \ge -q$ on R^2.

Remark 6. $\phi^1_{1,y} \ge 0$ on Ω^1_1, $\phi^2_{1,y} \ge 0$ on Ω^2_1. It follows from 4 and 5 above and known results (see Kindlehrer and Stampacchia) that the free surface $\partial\Omega^1_1 \cap R^1$ is analytic and monotone. If $y = y_{1,1}(x)$ is the free surface $\Gamma_{1,1}$ then

$$\phi^1_1(x, y_{1,1}(x)) = y_{1,1}(x) \qquad \text{on } \Omega^1_1 \cap R^1$$
$$\psi^1_1(x, y_{1,1}(x)) = 0 \qquad \text{on } \Omega^1_1 \cap R^1$$

Hence

$$\phi^1_{1,y} > 0 \qquad \text{on } \Omega^1_1 \cap R^1$$

Also $\phi^2_{1,y} > 0$ on $\Omega^2_1 \cap R^2$. Since $\phi^1_1 = 0$ on $y = 0$ and $\phi^1_1 \ge 0$ it follows that $\phi^3 \ge 0$ on $y = 0$. Also $\phi^1_{1,y} = \phi^3_{1,y} \ge 0$ on γ_1', hence $\phi^1_{1,y} \ge 0$ on Ω^1_1 by the minimum principle. Similarly for $\phi^2_{1,y}$.

8. Construct ψ^3_1 on R^3, the harmonic conjugate of ϕ^3_1 with $\psi^3_1(a_4) = 0$. Let

$$\omega^3_1(P) = \int_{a_4}^{P} (y - \phi^3_1)dy - \psi^3_1\, dx$$

Remark 7. At each stage of the iteration the regularity theorem insures that $\omega^1_n \in C^{1,\lambda}(\bar{R}^1)$, $\omega^2_n \in C^{1,\lambda}(\bar{R}^2)$.

Assumption 1. We shall assume that

$$\gamma_1 = \gamma_{1,n} \quad , \quad \gamma_2 = \gamma_{2,n} \quad \text{all } n$$

Assumption 2. $\gamma_1(\gamma_2)$ consists of two straight line segments (see Fig. 2) intersecting $y = 0$ at right angles.

Modified Scheme In place of step 6 set

$$\phi_n^3 = \phi_n^1 \text{ on the vertical portion of } \gamma_1$$

$$\phi_n^3 = \phi_n^2 \text{ on the vertical portion of } \gamma_2$$

$$\phi_n^3 = L \text{ on the non vertical portion of } \gamma_1, \gamma_2.$$

Let the projection of the non-vertical portion of γ_1, γ_2 onto γ_1', γ_2'. respectively, be of length $L > 1/k > 0$. Each iterate of the modified problem is a function of k, however this dependence will be omitted until the end of the discussion.

We shall treat the modified scheme together with the reflected scheme through $y = 0$.

Consider the reflected problems together with the original problems, for $y \leq 0$,

$\phi_n^j(x, -y) = -\phi_n^j(x,y)$, $j = 1,2,3$. Since $\phi_n^j(x,0) = 0$ let the reflection of γ be $\bar{\gamma}$ and $\gamma + \bar{\gamma} = \gamma^*$. Then

$$\phi_{n-1}^3(x_{14}, y) = \phi_n^1(x_{14}, y) \text{ for } -L \leq y \leq L \text{ i.e. on } \gamma_1'^*$$

$$\phi_{n-1}^3 = \phi_n^2 \text{ on } \gamma_2'^*$$

$$\phi_n^3 = \phi_n^1 \text{ on } \gamma_1^*$$

$$\phi_n^3 = \phi_n^2 \text{ on } \gamma_2^*$$

Hence

$$\phi_n^3 - \phi_{n-1}^3 = \phi_{n+1}^1 - \phi_n^1 \text{ on } \gamma_1'^*$$

$$\phi_n^3 - \phi_{n-1}^3 = \phi_{n+1}^2 - \phi_n^2 \text{ on } \gamma_2'^*$$

$$\phi^3_{n+1} - \phi^3_n = \phi^1_{n+1} - \phi^1_n \text{ on } \gamma_1^*$$

$$\phi^3_{n+1} - \phi^3_n = \phi^2_{n+1} - \phi^2_n \text{ on } \gamma_2^*$$

Note that $\phi^3_n - \phi^3_{n-1} = 0$ on $\{\gamma_1^* \cup \gamma_2^*\} \cap \{(x,y) \mid L - 1/k \leqq y \leqq L\} \cap$

$\{(x,y) \mid -L \leqq y \leqq -L + 1/k\}$

Remark 8. $\phi^3_{n,y} \geq 0$ on γ_1' and γ_2' Along $y = \alpha(x)$ and the non vertical parts of γ_1 and γ_2, $\phi^3_1 = L = \max_{R^3} \phi^3_n$. Along the vertical portions of γ_1, γ_2, $\phi^3_{1,y} = \phi^1_{1,y} \geq 0$. Also along $y = 0$, $\phi^3_{1,y} \geq 0$ since $\phi^3_1 \geq 0$ on R^3 and $\phi^3_1 = 0$ on $y = 0$. By the minimum principle the result follows for $\phi^3_{1,y}$.

Making use of the previous remarks gives the general result.

Remark 9. $y - \phi^3_n \leq 0$ on R^3, $L \geq \phi^3_n \geq 0$ on R^3, $\omega^1_{n,y} \leq 0$ on R^1, $\omega^2_{n,y} \leq 0$ on R^2, $\omega^1_{n,x} \leq 0$ on R^1, $\omega^2_{n,x} \geq -q$ on R^2. These follow using the previous arguments.

Remark 10. The free surfaces $\partial\Omega^1_n \cap R^1$ and $\partial\Omega^2_n \cap R^2$ are analytic monotone curves. This follows from Remark 9 and Remark 6.

7. BASIC LEMMAS

The following lemmas are essential for the study of convergence.

Lemma 1. Let v be harmonic in the region G continuous on $\bar{G}$ bounded by the polygon (see Fig. 3). Assume $v = 0$ on a^* and $0 \leqq |v| \leqq 1$ on $b^* = b_1^* + b_1'^*$. Then there exists a positive constant $q_1 < 1$ independent of v such that $|v| \leq q_1$ on all of b'^*.

Lemma 2. Let v be harmonic in the region G continuous on $\bar{G}$ bounded by the curvilinear polygon (see Fig. 4). Assume $v = 0$ on $a^* + a_3^*$ and $0 \leq |v| \leq 1$ on $b^* = b_1^* + b_2^*$ then there exists a positive constant $q_2 < 1$ independent of v such that

$$|v| \leq q_2 \qquad \text{on all of } b'^*$$

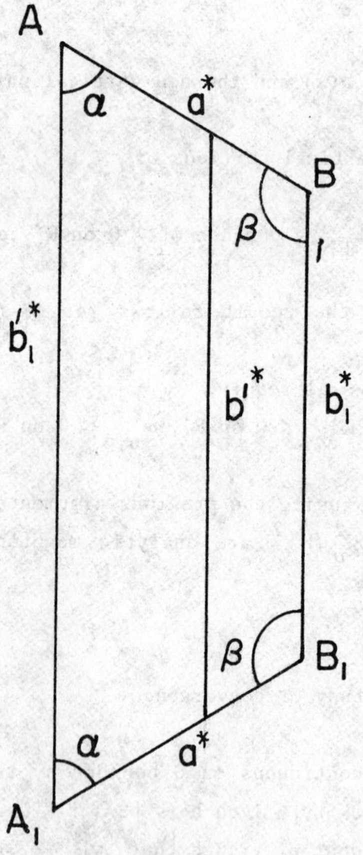

Fig. 3

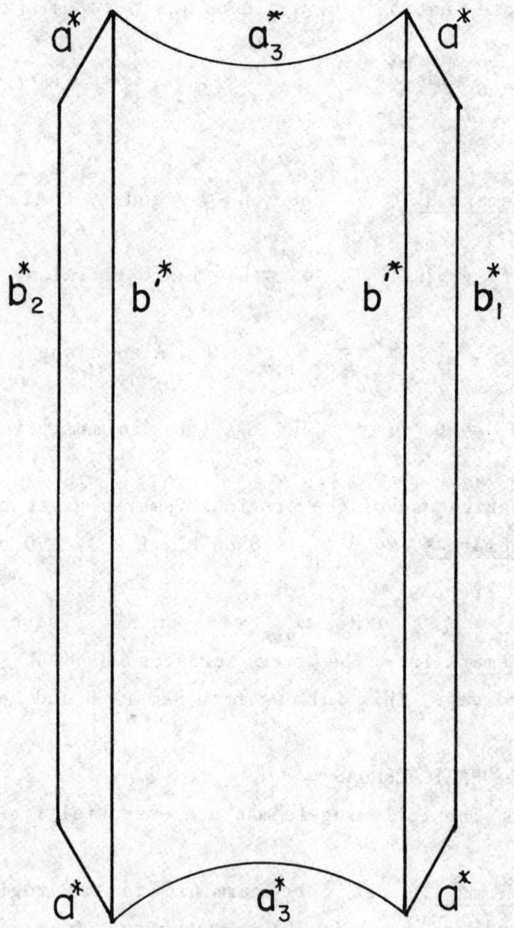

Fig. 4

Proof of 1. Consider

$$W_1(P) = \int_A^{A_1} \frac{\partial \log}{\partial \nu_Q} (1/r)\, ds_Q \quad \text{Along } b_1'^*, \quad r = |P - Q|, \quad Q \in b_1'^*, \quad P \notin b_1'^*$$

$$W_2(P) = \int_{B_1}^{B} \frac{\partial \log}{\partial \nu_Q} (1/r)\, ds_Q \quad \text{Along } b_1^*, \quad P \notin b_1^*$$

$W_1(P)$ = angle that radius vector centered at P makes in moving from A to A_1 along $b_1'^*$. Similarly for $W_2(P)$. $W_1(P)$ is harmonic in G and continuous on the open arc $b_1'^*$. A similar remark holds for $W_2(P)$ and b_1^*. Note that:

$$\underset{\substack{P \to A \\ P \in b_1'^*}}{W_1(P) \to R_{A+}}\,; \qquad \underset{\substack{P \to A \\ P \in a^*}}{W_1(P) \to R_{A-}}\,, \qquad R_{A+} - R_{A-} = \alpha$$

$$\underset{\substack{P \to A_1 \\ P \in b_1'^*}}{W_1(P) \to R_{A_1^+}}\,; \qquad \underset{\substack{P \to A_1 \\ P \in a^*}}{W_1(P) \to R_{A_1^-}}\,, \qquad R_{A_1^+} - R_{A_1^-} = \alpha$$

Also note

$$\underset{\substack{P \to B \\ P \in b_1^*}}{W_2(P) \to R_{B+}}\,; \qquad \underset{\substack{P \to B \\ P \in a^*}}{W_2(P) \to R_{B-}}\,, \qquad R_{B+} - R_{B-} = \beta$$

$$\underset{\substack{P \to B_1 \\ P \in b_1^*}}{W_2(P) \to R_{B_1^+}}\,; \qquad \underset{\substack{P \to B \\ P \in a^*}}{W_2(P) \to R_{B_1^-}}\,, \qquad R_{B_1^+} - R_{B_1^-} = \beta$$

Let ℓ_δ be any straight line in G meeting A at an angle δ with $b_1'^*$, $0 < \delta < \alpha$. Then

$$\underset{\substack{P \to A \\ P \in \ell_\delta}}{W_1(P) \to R_{A-} + \delta} = \frac{\delta}{\alpha} R_{A+} + (1 - \delta/\alpha) R_{A-}$$

Similarly if $\ell_{1\delta}$ makes an angle δ with $b_1'^*$ at A_1

$$\underset{\substack{P \to A \\ P \in \ell_{1\delta}}}{W_1(P) \to R_{A_1^-} + \delta} = \frac{\delta}{\alpha} R_{A+} + (1 - \delta/\alpha) R_{A-}$$

Similarly if $\ell_{\overline{\delta}}$, $\ell_{1\overline{\delta}}$ makes an angle $\overline{\delta}$, $0 < \overline{\delta} < \beta$, with a^* at B and B_1, respectively, then

$$W_2(P) \to R_{B^-} + \overline{\delta} = \frac{\overline{\delta}}{\beta} R_{B^+} + (1 - \frac{\overline{\delta}}{\beta})R_{B^-}$$
$$P \to B$$
$$P \in \ell_{\overline{\delta}}$$
$$W_2(P) \to R_{B_1^-} + \overline{\delta} = \frac{\overline{\delta}}{\beta} R_{B_1^+} + (1 - \frac{\overline{\delta}}{\beta})R_{B_1^-}$$
$$P \to B_1$$
$$P \in \delta_{1\overline{\delta}}$$

Consider the functions $\rho_{b_1^*}$ and $\rho_{b_1'^*}$ on $a^* + b_1^* + b_1'^*$ with

$$\rho_{b_1'^*} = \alpha \text{ on } b_1'^*, \qquad \rho_{b_1^*} = \beta \text{ on } b_1^*$$
$$= 0 \text{ on } a^* + b_1^* \qquad = 0 \text{ on } a^* + b_1'^*$$

Let $\overline{w}_1$ and $\overline{w}_2$ be the respective boundary values of W_1 and W_2 on $a^* + b_1'^* + b_1^*$. Then $\overline{w}_1 - \rho_{b_1'^*}$ and $\overline{w}_2 - \rho_{b_1^*}$ are continuous on $a^* + b_1'^* + b_1^*$. Let F_1 and F_2 be harmonic functions on G with boundary values $\overline{w}_1 - \rho_{b_1'^*}$ and $\overline{w}_2 - \rho_{b_1^*}$ respectively.

Consider

$$S_1(P) = (W_1 - F_1)/\alpha$$
$$S_2(P) = (W_2 - F_2)/\beta$$

$S_1(P)$ and $S_2(P)$ are harmonic in G with

$$S_1 = 1 \text{ on } b_1'^*, \qquad S_2 = 0 \text{ on } a^* + b'^*_1$$
$$= 0 \text{ on } a^* + b_1^* \qquad = 1 \text{ on } b_1^*$$

Hence if $S = S_1 + S_2$, S is harmonic on G, bounded on $\overline{G}$

$$S = 1 \text{ on } b_1'^* + b_1^*$$
$$= 0 \text{ on } a^*$$

Also as approach A, A_1, B or B_1 from interior the limit of S must lie between 0 and 1.

If we restrict ourselves to points on the line b'^* including end points, then there exists a $q_1 < 1$ such that $S(P) \leq q_1 < 1$ for a P on b'. Since if not, then there exists a sequence of P_n on b'^* such that

$$\lim_{n\to\infty} S(P_n) = 1$$

Since b'^* with boundary points is closed, from this it follows:

$$S(P_o) = 1, \quad P_o \in b'^*$$

P_o can not lie on a^* since $S(P) = 0$ on a^*. Moreover P_o can not be an interior point of b' by the maximum principle.

Consider $S - v$

$$\begin{aligned} S - v &= 0 \text{ on } a^* \\ &\geq 0 \text{ on } b_1^* + b_1'^* \\ &= \text{same value as } S \text{ as approach } A, A_1, B, B_1 \end{aligned}$$

Hence

$$S - v \geq 0 \text{ in } \bar{G}$$

Similarly $\quad S + v \geq 0$ in $\bar{G}$

Thus

$$-q_1 \leq -S \leq v \leq S \leq q_1 \text{ on } b'^*$$

and the proof is complete.

The proof of Lemma 2 follows same lines.

8. CONVERGENCE OF THE MODIFIED SCHEME

We first note that $\phi^1_{n+1} - \phi^1_n$ can not take on a positive maximum on $\partial\Omega^1_n \cap R^1$. This follows from the maximum principle, as in the uniqueness theorem, since $\phi^1_{n+1} - \phi^1_n$ is:

(1) harmonic on $\Omega^1_{n+1} \cap \Omega^1_n$,

(2) harmonic on $R^1 - \overline{(\Omega^1_{n+1} \cap \Omega^1_n)}$,

(3) in $C^\lambda(R^1)$,

(4) $= 0$ at a_4.

(5) $= 0$ on $\overline{(R^1 - \Omega^1_n)} \cap \overline{(R^1 - \Omega^1_{n+1})}$

(6) $\partial\Omega^1_n \cap R^1$, $\partial\Omega^1_{n+1} \cap R^1$ have continuous tangent

Similarly $\phi^1_{n+1} - \phi^1_n$ can not take on a negative minimum on $\partial\Omega^1_n \cap R_1$. Since $\phi^1_{n+1} - \phi^1_n = 0$ on $y = 0$ it follows

$$\max_{R^1} | \phi^1_{n+1} - \phi^1_n | \leq \max_{\gamma'_1} | \phi^1_{n+1} - \phi^1_n |$$

where the inequality follows from the maximum principle. It follow that:

$$\max_{R^{1*}} | \phi^1_{n+1} - \phi^1_n | \leq \max_{\gamma'^*_1} | \phi^1_{n+1} - \phi^1_n |$$

where we have set

$$R^{1*} = R^1 \cup \bar{R}^1 \quad , \quad (\bar{R}^1 = \text{reflection of } R^1)$$

Similarly

$$\max_{R^{2*}} |\phi^2_{n+1} - \phi^2_n| \le \max_{\gamma_2'^*} |\phi^2_{n+1} - \phi^2_n|$$

Let

$$M_2^{1'} = \max_{\gamma_1'^*} |\phi^1_{n+1} - \phi^1_n|, \quad M_n^{2'} = \max_{\gamma_2'^*} |\phi^2_{n+1} - \phi^2_n|$$

$$M_n^3 = \max_{\gamma_1^* \cup \gamma_2^*} |\phi^3_{n+1} - \phi^3_n|$$

$$v_n^1 = \frac{\phi^1_{n+1} - \phi^1_n}{M_n^{1'}}, \quad v_n^2 = \frac{\phi^2_{n+1} - \phi^2_n}{M_n^{2'}}, \quad v_n^3 = \frac{\phi^3_{n+1} - \phi^3_n}{M_n^3}$$

Let a_1^* be the non vertical portion of γ_1 together with its mirror image. Similarly for a_2^* and γ_2. Let a_3^* be the curve $y = a(x)$, $x_{24} \le x \le x_{14}$, and its mirror image.

Note that $|v_n^1| \le 1$ on R^{1*}, $\quad v_n^1 = 0$ on a_1^*

$|v_n^2| \le 1$ on R^{2*}, $\quad v_n^2 = 0$ on a_2^*

$|v_n^3| \le 1$ on R^{3*}, $\quad v_n^3 = 0$ on a_3^*

In Lemma 1 take $b'^* = \gamma_1^*$, $b_1'^* = \gamma_1'^*$, b_1^* to right of the vertical portion of γ_1^*, as shown in figure 3. Similarly for γ_2^* and $\gamma_2'^*$. Then by Lemma 1

$$|v_n^1| \le q_1 \text{ on } \gamma_1^*, \quad |v_n^2| \le q_1 \text{ on } \gamma_2^*$$

In Lemma 2 take $b^* = \gamma_1^* \cup \gamma_2^*$, $b_1^* = \gamma_1^*$, $b_2^* = \gamma_2^*$ then

$$|v_n^3| \le q_2 \text{ on } \gamma_1'^* \cup \gamma_2'^*$$

Let $\bar{q} = \min\{q_1, q_2\}$, then if we set

$$M_n^{11} = \max_{\gamma_1^*} |\phi^1_{n+1} - \phi^1_n|, \quad M_n^{13} = \max_{\gamma_1^*} |\phi^3_{n+1} - \phi^3_n|$$

$$M_n^{22} = \max_{\gamma_2^*} |\phi^2_{n+1} - \phi^2_n|, \quad M_n^{23} = \max_{\gamma_2^*} |\phi^3_{n+1} - \phi^3_n|$$

$$M_n^{11'} = \max_{\gamma_1'^*} |\phi^1_{n+1} - \phi^1_n|, \quad M_n^{13'} = \max_{\gamma_1'^*} |\phi^3_{n+1} - \phi^3_n|$$

$$M_n^{22'} = \max_{\gamma_2'^*} |\phi^2_{n+1} - \phi^2_n|, \quad M_n^{23'} = \max_{\gamma_2'^*} |\phi^3_{n+1} - \phi^3_n|$$

It follows that

$$M_n^{11} = M_n^{13} \le M_n^3\ ,\ M_n^{22} = M_n^{23} \le M_n^3\ ,\ M_n^{11'} = M_{n-1}^{13'} = M_n^{1'},\ M_n^{22'} = M_{n-1}^{23'} = M_n^{2'}$$

By the above

$$M_n^{11} \le \bar{q}\, M_n^{1'} \qquad\qquad M_n^{22} \le \bar{q}\, M_n^{2'}$$

$$M_n^{13'} \le \bar{q}\, M_n^{3} \qquad\qquad M_n^{23'} \le \bar{q}\, M_n^{3}$$

Hence

$$M_n^{11} \le \bar{q}\, M_{n-1}^{13'} \le \bar{q}^2\, M_{n-1}^3\ , \qquad M_n^{22} \le \bar{q}^2\, M_{n-1}^3$$

$$\max(M_n^{11}, M_n^{22}) \le \bar{q}^2\, M_{n-1}^3 = \bar{q}^2 \max(M_{n-1}^{13}, M_{n-1}^{23}) = \bar{q}^2 \max(M_{n-1}^{11}, M_{n-1}^{22})$$

i.e.

$$\tilde{M}_n \le \bar{q}^2\, \tilde{M}_{n-1}$$

where

$$\tilde{M}_n = \max(M_n^{11}, M_n^{22}) = \max(M_n^{13}, M_n^{23})$$

Hence $\tilde{M}_n \to 0$ as $n \to \infty$.

Also

$$\max(M_n^{13'}, M_n^{23'}) \le \bar{q} \max(M_n^{13}, M_n^{23}) = \bar{q} \max(M_n^{11}, M_n^{22}) \le \bar{q}^2 \max(M_n^{1'}, M_n^{2'})$$

$$\le \bar{q}^2 \max(M_n^{13'}, M_n^{23'})$$

i.e.

$$\tilde{M}_n^3 \le \bar{q}^2\, \tilde{M}_{n-1}^3$$

where

$$\tilde{M}_n^3 = \max(M_{n-1}^{13'}, M_{n-1}^{23'}) = \max(M_n^{11'}, M_n^{22'})$$

Hence $\tilde{M}_n^3 \to 0$ as $n \to \infty$.

The following converges uniformly on $\gamma_1'^{*} \cup \gamma_2'^{*} \cup \gamma_1^{*} \cup \gamma_2^{*}$

$$\phi_o^3 + \sum_{n=0}^{\infty} (\phi_{n+1}^3 - \phi_n^3) = \lim_{n\to\infty} \phi_n^3 = \phi^3$$

since it is majorized by a geometric series. Similarly the sequences

$$\phi_n^1 \to \phi^1 \quad \text{as } n \to \infty$$

$$\phi_n^2 \to \phi^2 \quad \text{as } n \to \infty$$

converge uniformly on $\gamma_1'^* \cup \gamma_1^*$ and on $\gamma_2'^* \cup \gamma_2^*$, respectively.

Note that

$$\phi_n^3 - \phi_n^1 = 0 \text{ on } \gamma_1^*$$

$$\phi_n^3 - \phi_n^2 = 0 \text{ on } \gamma_2^*$$

$$\phi_n^3 - \phi_n^1 = \phi_n^3 - \phi_{n-1}^3 \to 0 \text{ uniformly on } \gamma_1'^*$$

$$\phi_n^3 - \phi_n^2 = \phi_n^3 - \phi_{n-1}^3 \to 0 \text{ uniformly on } \gamma_2'^*$$

Thus $\phi^3 = \phi^1$ on $\gamma_1^* \cup \gamma_1'^*$, $\phi^3 = \phi^2$ on $\gamma_2^* \cup \gamma_2'^*$ and hence

$\phi^3 = \phi^1 =$ harmonic function in the region bounded by $\gamma_1'^* \cup \gamma_1^*$.

Similarly for $\phi^3 = \phi^2$ on the region bounded by $\gamma_2'^* \cup \gamma_2^*$.

Thus if

$$\begin{aligned} \phi &= \phi^3 \text{ in } R^3 \\ &= \phi^1 \text{ in the region bounded by } \gamma_1'^* \cup \gamma_1^* \\ &= \phi^2 \text{ in the region bounded by } \gamma_2'^* \cup \gamma_2^* \end{aligned}$$

Then ϕ is harmonic in R^3. In order to harmonically continue ϕ up to the free boundaries Γ_1 and Γ_2 we solve the corresponding variational inequalities with

$$\omega^1 = \int_{\gamma_1'} (y - \phi^1)dy \quad , \quad \omega^2 = \int_{\gamma_2'} (y - \phi^2)dy$$

and note that on the overlap bounded by $\gamma_1'^* \cup \gamma_1^*$, $w_y^1 = y - \phi^3$ and on the overlap bounded by $\gamma_2'^* \cup \gamma_2^*$, $w_y^2 = y - \phi^3$. From the solution of the variational inequalities it follows for the harmonic conjugate ψ of ϕ that $\psi = 0$ on Γ_1 and $\psi = q$ on Γ_2. Also from the variational inequality it follows that ϕ and $\psi \in C^\lambda$ on the region bounded by $y = \alpha(x)$, $y = 0$, Γ_1 and Γ_2. Also since

$$\omega_y^1 = y - \phi^1 = \lim_{n\to\infty} (y - \phi_n^1) = \lim_{n\to\infty} \omega_{n,y}^1 \leq 0$$

$$\omega_x^1 = \lim_{n\to\infty} w_{n,x}^1 \leq 0 \quad (\text{since } \phi_n^1 \text{ converges implies } \psi_n^1 \text{ converges implies } \omega_{n,x}^1$$

converges).

Similarly $w_y^2 \leq 0$, $w_x^2 \geq -q$. Hence Γ_1 and Γ_2 are monotone analytic curves. q can be obtained by integrating the Cauchy Riemann equation along $y = 0$.

Remark 1. The solutions above depend on k. To emphasize this denote ϕ by $\phi(k)$. Note that $0 \leq \phi(k) \leq L$ for all k. Hence by the compactness of harmonic functions it follows there exists a subsequence $\phi(k_j)$ that coverges to ϕ harmonic in Ω satisfying the appropriate boundary conditions. Moreover $\Gamma_1(k_j)$ and $\Gamma_2(k_j)$ converge to the free surfaces Γ_1, Γ_2, respectively, for the original problem.

9. NUMERICAL RESULTS

The following test problem was used to initially test the proposed scheme. The problem is that of seepage from a trapezoidal channel into permeable soil underlain at a finite depth by a drain. This problem is solved by Bruch et. al. [1] using the standard Baiocchi method and transformation on only half the channel since it was a symmetric channel. The channel had a side slope of 45° and the rest of the channel dimensions are given in [1]. They solved the problem using two mesh spacings, $\Delta x = \Delta y = 1$ft and $\Delta x = \Delta y = 2$ft.

Even though this case was a symmetric shaped channel, it was felt that the results would be representative of the proposed scheme applied to such seepage problems. The proposed scheme was applied with a $\Delta x = 1$ft and $\Delta y = 2$ft for the mesh spacing and gave the following results. The free surface location almost exactly matched the results given by. [1] in their fig. 2. It took 53 iterations and a computer time of .87 seconds on an Itel AS/6 (which does not include the compilation or linkage editing times) to obtain this shape. It should be noted that in the numerical algorithm, the parameters q and A do not affect the results significantly. Thus the necessity of outer iterations to obtain these values in the scheme is unnecessary.In [1] the flowrate, q, had to be determined as part of the solution and thus the number of iterations and computer execution time was considerably greater. With the proposed scheme, if the flowrate were desired, it would be necessary to solve for ϕ in the ω part of the flowfield on the left and right and then use the Cauchy-Riemann equations to solve for ψ and thus obtain the flowrate.

ACKNOWLEDGEMENT

The numerical computations were funded by a University of California at Santa Barbara Academic Senate Computer Research Grant.

REFERENCES

1. Bruch, J.C., Jr., Sayle, F.C., and Sloss, J.M., "Seepage from a Trapezoidal and a Rectangular Channel Using Variational Inequalities," J. Hydrology, 36, 247-260 (1978).
2. Courant, R. and Hilbert, D., Methods of Mathematical Physics, Vol. II, Partial Differential Equations by R. Courant, Interscience Publishers, New York (1962).
3. Kinderlehrer, D. and Stampacchia, G., An Introduction to Variational Inequalities and Their Applications, Academic Press, New York (1980).
4. Remar, J., Bruch, J.C., Jr. and Sloss, J.M., "Axisymmetric Free Surface Seepage," Num. Math., submitted.

Contemporary Mathematics
Volume 11, 1982

Bers-Vekua Equations of Two Complex Variables

J. L. Buchanan*
Department of Mathematics
U. S. Naval Academy
Annapolis, MD 21402

§1. Introduction

Solutions to the Bers-Vekua equation

(1.1) $\partial_{\bar{z}} w(z,\bar{z}) = A(z,\bar{z})w(z,\bar{z}) + B(z,\bar{z})\overline{w}(z,\bar{z})$ ($\partial_{\bar{z}} := 1/2(\partial_x + i\partial_y)$) have been comprehensively studied and are known to possess many properties of analytic functions [1], [8], [9]. More recently attention has been given to the several complex variable analogue of equation (1.1)

(1.2) $\partial_{\bar{z}_j} w(z,\bar{z}) = A_j(z,\bar{z})w(z,\bar{z}) + B_j(z,\bar{z})\overline{w}(z,\bar{z})$, $j=1,\ldots,n$, $(z:=(z_1,\ldots,z_n))$.

Various aspects of this system (1.2) have been studied by Koohara [2], [3] and Tutschke [4], [5], [6], [7]. The system (1.2) is in general overdetermined and thus compatibility conditions on the coefficients A_j and B_j are necessary for there to be solutions at all. One such set of conditions is given by Koohara in [2]. In this paper we seek, under certain nonsigularity assumptions, all sets of compatibility conditions which permit solution of the simplest system for n = 2

$$\partial_{\bar{z}_j} w = B_j \overline{w}\ , \quad j = 1,2. \tag{1.3}$$

The system (1.2) can be reduced to (1.3) when the A_j and B_j satisfy certain relations [2].

Since our goal is to be comprehensive in the sense of cataloguing all sets of compatibility conditions permitting (1.3) to have solutions, we, in addition to restricting attention to two complex variables, make the further simplifying assumption that the B_j are analytic functions of their real arguments (x_1, y_1, x_2, y_2) and seek solutions which are analytic in a neighborhood of a

*This research is funded by the Naval Academy Research Council.

fixed given point $(x_1{}^o, y_1{}^o, x_2{}^o, y_2{}^o)$. The assumption of analyticity permits us to solve, instead of (1.3), the system

$$(1.4) \qquad \partial_{\zeta_j} w(z,\zeta) = B_j(z,\zeta)w^*(\zeta,z) \qquad j = 1,2,$$

$(\zeta:=(\zeta_1,\zeta_2),\ z_j = x_j + iy_j,\ \zeta_j = x_j - iy_j;\ x_j, y_j \in C)$ where B_j and w are now analytic functions of the four complex variables $(z_1, z_2, \zeta_1, \zeta_2)$ and

$$f^*(\zeta,z) := \overline{f}(\overline{\zeta}_1, \overline{\zeta}_2, \overline{z}_1, \overline{z}_2).$$

It is easily seen by power series arguments that $\partial_{\zeta_j} w(z,\overline{z}) = \partial_{\overline{z}_j} w(z,\overline{z})$ and $w^*(\overline{z},z) = \overline{w}(z,\overline{z})$ when the x_j and y_j are restricted to real values. Thus we may solve (1.3) by solving (1.4) for $w(z,\zeta)$ and setting $\zeta = \overline{z}$. Henceforth we denote the point about which we seek a solution by (z^o, ζ^o) with $\zeta^o = \overline{z}^o$.

Following Koohara [2] we seek a change of coordinates which will make the system (1.4) more tractable. We impose at the outset the nonsingularity assumption

$$(1.5) \qquad B_2(z^o, \zeta^o) \neq 0.$$

We look for a new set of coordinates in the form

$$(1.6) \quad \xi_1 = z_1,\ \xi_2 = \xi_2(z,\zeta)\ ,\ \eta_1 = \zeta_1\ ,\ \eta_2 = \eta_2(z,\zeta).$$

The Jacobian of this transformation is

$$J[\xi,\eta](z,\zeta) := \partial_{z_2}\xi_2\partial_{\zeta_2}\eta_2 - \partial_{z_2}\eta_2\partial_{\zeta_2}\xi_2.$$

Hence we require $J[\xi,\eta](z^o,\zeta^o) \neq 0$.

Define

$$(1.7) \qquad B = -\frac{B_1}{B_2}.$$

Under the proposed coordinate change our system has the form

$$\partial_{\eta_1} w + \partial_{\zeta_1}\xi_2\partial_{\xi_2} w + \partial_{\zeta_1}\eta_2\partial_{\eta_2} w = B_1 w^*$$

$$-B\partial_{\zeta_2}\xi_2\partial_{\xi_2} w - B\partial_{\zeta_2}\eta_2\partial_{\eta_2} w = B_1 w^*.$$

Subtracting the second equation from the first, we obtain

$$(1.8) \quad \partial_{\eta_1} w + (\partial_{\zeta_1}\xi_2 + B\partial_{\zeta_2}\xi_2)\partial_{\xi_2} w + (\partial_{\zeta_1}\eta_2 + B\partial_{\zeta_2}\eta_2)\partial_{\eta_2} w = 0.$$

We may solve the partial differential equation

$$(1.9) \qquad \partial_{\zeta_1}\lambda + B\partial_{\zeta_2}\lambda = 0$$

by the method of characteristics. The differential equation

$$\frac{d\zeta_2}{d\zeta_1} = B(z,\zeta)$$

has general solution $\zeta_2 = \zeta_2(z_1, z_2, \zeta_1, \alpha)$ where $\partial_\alpha \zeta_2 \neq 0$ in a neighborhood of (z^o, ζ^o). Thus we may solve for α as a function of (z,ζ) with $\partial_{\zeta_2} \alpha(z^o, \zeta^o) \neq 0$. If we introduce the coordinates $(\hat{z}_1, \hat{z}_2, \hat{\zeta}_1, \hat{\zeta}_2) = (z_1, z_2, \zeta_1, \alpha)$, then all solutions to (1.9) have the form $\lambda = \lambda(\hat{z}_1, \hat{z}_2, \hat{\zeta}_2)$. If (ξ,η) is a change of coordinates of the form (1.6) for which ξ_2 and η_2 are both solutions of (1.9), then from (1.8) $\partial_{\eta_1} w \equiv 0$. Henceforth we assume that ξ_2 and η_2 satisfy (1.9) for any coordinate change (ξ,η) under consideration.

In the coordinates (ξ,η) the system (1.4) is equivalent to a single independent equation

(1.10) $$\partial_{\eta_2} w + \gamma \partial_{\xi_2} w = Cw^*$$

where

$$\gamma := \frac{\partial_{\zeta_2} \xi_2}{\partial_{\zeta_2} \eta_2} \quad , \quad C := \frac{B_2}{\partial_{\zeta_2} \eta_2}$$

and we have assumed, without loss of generality, that $\partial_{\zeta_2} \eta_2(z^o, \zeta^o) \neq 0$. Note that from (1.9)

(1.11a) $$\partial_{\eta_1} \zeta_2 = \frac{-\partial_{z_2} \xi_2 \partial_{\zeta_1} \eta_2 + \partial_{z_2} \eta_2 \partial_{\zeta_1} \xi_2}{J[\xi_2, \eta_2]} = B$$

and

(1.11b) $$\partial_{\eta_1} z_2 = 0 .$$

Thus

(1.12) $$\partial_{\eta_1} \gamma = 0 .$$

In obtaining (1.12) we have used the identity

(1.13) $$\partial_{\zeta_1 \zeta_2} \lambda + B \partial_{\zeta_1 \zeta_2} \lambda = - \partial_{\zeta_2} B \partial_{\zeta_2} \lambda$$

which follows from (1.9).

From (1.10) and $\partial_{\eta_1} w = 0$ we have

$$\partial_{\eta_1} (Cw^*) = 0 ,$$

and hence from (1.11)

$$(\partial_{\eta_1} C)w^* + C(\partial_{\zeta_1} w^* + B\partial_{\zeta_2} w^*) = 0 .$$

By the <u>involution</u> of a function $f(z,\zeta)$ we shall mean $f^*(\zeta,z)$. Taking the involution of the above equation and then changing to (ξ,η)-coordinates yields the auxillary equation

(1.14) $$\partial_{\xi_1} w + (\partial_{z_1}\xi_2 + B^*\partial_{z_2}\xi_2)\partial_{\xi_2} w + (\partial_{z_1}\eta_2 + B^*\partial_{z_2}\eta_2)\partial_{\eta_2} w = -(\partial_{\eta_1}\log C)^* w$$

We regard the involution as taken with respect to the original variables (z,ζ), unless otherwise indicated.

§2. <u>Three regular cases</u>

In this section we work in the coordinates $(\hat{z}_1, \hat{z}_2, \hat{\zeta}_1, \hat{\zeta}_2)$ defined in connection with equation (1.9). Equation (1.14) becomes

(2.1) $$\partial_{\hat{z}_1} w + B^*\partial_{\hat{z}_2} w + (\partial_{z_1}\hat{\zeta}_2 + B^*\partial_{z_2}\hat{\zeta}_2)\partial_{\hat{\zeta}_2} w = -(\partial_{\hat{\zeta}_1}\log\hat{C})^* w.$$

Since $\partial_{\eta_1} w = 0$, we have upon differentiation

(2.2) $$\partial_{\hat{\zeta}_1} B^*\partial_{\hat{z}_2} w + \partial_{\hat{\zeta}_1}(\partial_{z_1}\hat{\zeta}_2 + B^*\partial_{z_2}\hat{\zeta}_2)\partial_{\hat{\zeta}_2} w = -\partial_{\hat{\zeta}_1}(\partial_{\hat{\zeta}_1}\log\hat{C})^* w.$$

We take as the defining condition of the first case

<u>Case I</u>: $\partial_{\hat{\zeta}_1} B^* = \partial_{\zeta_1} B^* + B\partial_{\zeta_2} B^* \equiv 0$.

Since $\hat{\zeta}_2$ satisfies (1.9), it is not difficult to see that

(2.3) $$\begin{aligned}\partial_{\hat{\zeta}_1}&(\partial_{z_1}\hat{\zeta}_2 + B^*\partial_{z_2}\hat{\zeta}_2)\\ &= (\partial_{\zeta_1} B^* + B\partial_{\zeta_2} B^*)\partial_{z_2}\hat{\zeta}_2 - (\partial_{\zeta_1} B^* + B\partial_{\zeta_2} B^*)^*\partial_{\zeta_2}\hat{\zeta}_2\\ &= (\partial_{\hat{\zeta}_1} B^*)\partial_{z_2}\hat{\zeta}_2 - (\partial_{\hat{\zeta}_1} B^*)^*\partial_{\zeta_2}\hat{\zeta}_2 .\end{aligned}$$

Hence in Case I the coefficient of $\partial_{\hat{\zeta}_2} w$ also vanishes.

Now suppose that $\partial_{\hat{\zeta}_1} B^* \not\equiv 0$. We impose the nonsingularity assumption

(2.4) $$\partial_{\hat{\zeta}_1}(\partial_{z_1}\hat{\zeta}_1 + B^*\partial_{z_2}\hat{\zeta}_2) = D_1(\hat{z},\hat{\zeta})\partial_{\hat{\zeta}_1} B^*$$

for some function D_1 regular in a neighborhood of $(\hat{z}^o,\hat{\zeta}^o):=(\hat{z}(z^o,\zeta^o),\hat{\zeta}(z^o,\zeta^o))$. Since $\partial_{\hat{\zeta}_1} = \partial_{\zeta_1} + B\partial_{\zeta_2}$, the defining condition of Case I can be written in terms of derivatives with respect to the original coordinates $(\hat{z},\hat{\zeta})$. As the defining condition of the second case we take

Case II: $\quad \partial_{\hat{\zeta}_1} D_1 = \partial_{\zeta_1} D_1 + B\partial_{\zeta_2} D_1 \equiv 0$.

Hence equation (2.2) may be written

$$\partial_{\hat{z}_2} w + D_1 \partial_{\hat{\zeta}_2} w = E_1 w \tag{2.5}$$

where we assume that there is a function E_1, regular in a neighborhood of $(\hat{z}^0, \hat{\zeta}^0)$ such that

$$-\partial_{\hat{\zeta}_1} (\partial_{\hat{\zeta}_1} \log \hat{C})^* = E_1 \partial_{\hat{\zeta}_1} B^*. \tag{2.6}$$

Finally suppose $\partial_{\hat{\zeta}_1} D_1 \not\equiv 0$. Let us assume that there is a regular function F_1 in a neighborhood of $(\hat{z}^0, \hat{\zeta}^0)$ such that

$$\partial_{\hat{\zeta}_1} E_1 = F_1 \partial_{\hat{\zeta}_1} D_1 \tag{2.7}$$

As the defining condition for Case III we take

Case III: $\quad \partial_{\hat{\zeta}_1} F_1 = \partial_{\zeta_1} F_1 + B\partial_{\zeta_2} F_1 \equiv 0$.

In Case II and Case III it is possible to write the defining conditions in terms of derivatives with respect to the original variables. Hence the conditions are independent of the particular choice of $\hat{\zeta}_2$. Note that except for instances when (z^0, ζ^0) lies on the singular manifold of one of the functions B, D_1, E_1, or F_1 there can be no nontrivial solutions to the original system (1.4) in which one of the three conditions does not obtain.

§3. Solution of Case I.

Lemma (3.1) (c.f. Koohara [2]): If $(\partial_{\zeta_1} + B\partial_{\zeta_2})B^* = 0$, then we may find a nonsingular change of coordinates (ξ, η) such that ξ_2 and η_2 both satisfy the system

$$(\partial_{\zeta_1} + B\partial_{\zeta_2})\lambda = 0 \tag{3.1a}$$

$$(\partial_{z_1} + B^*\partial_{z_2})\lambda = 0 . \tag{3.1b}$$

Moreover ξ_2 and η_2 may be chosen so that $\xi_2^*(\zeta, z) = \eta_2(z, \zeta)$.

Proof: As noted previously (see (1.9)) all solutions to (3.1a) have the form $\lambda = \lambda(\hat{z}_1, \hat{z}_2, \hat{\zeta}_2)$. Equation (3.1b) then implies

(3.2) $\partial_{\hat\zeta_1}\lambda + B^*\partial_{\hat\zeta_2}\lambda + (\partial_{z_1}\hat\zeta_2 + B^*\partial_{z_2}\hat\zeta_2)\partial_{\hat\zeta_2}\lambda = 0$.

From (2.3), the hypothesis of the Lemma implies that B^* and $(\partial_{z_1}\hat\zeta_2 + B^*\partial_{z_2}\hat\zeta_2)$ are functions of $\hat z_1$, $\hat z_2$, and $\hat\zeta_2$ alone. Consequently we may solve the system of ordinary differential equations

$$\frac{d\hat\zeta_2}{d\hat z_1} = B^* \qquad \frac{d\hat\zeta_2}{d\hat z_1} = \partial_{z_1}\hat\zeta_2 + B^*\partial_{z_2}\hat\zeta_2$$

to obtain a family of solutions

$$\hat z_2 = \hat z_2(\hat z_1, \beta_1, \beta_2) \quad , \quad \hat\zeta_2 = \hat\zeta_2(\hat z_1, \beta_1, \beta_2)$$

which may be inverted to yield functions $\beta_j = \beta_j(\hat z_1, \hat z_2, \hat\zeta_2)$ satisfying $\partial_{\hat z_2}\beta_1\partial_{\hat\zeta_2}\beta_2 - \partial_{\hat\zeta_2}\beta_1\partial_{\hat z_2}\beta_2 \neq 0$ at $(\hat z_1{}^o, \hat z_2{}^o, \hat\zeta_2{}^o) := (z_1{}^o, z_2{}^o, \zeta_2(z^o,\zeta^o))$. In terms of the original variables (z,ζ) this condition is

(3.3) $$\frac{1}{\partial_{\zeta_2}\hat\eta_2}(\partial_{z_2}\beta_1\partial_{\zeta_2}\beta_2 - \partial_{z_2}\beta_2\partial_{\zeta_2}\beta_1) \neq 0$$

at (z^o, ζ^o). From (3.2) we see that any function $\lambda = \lambda(\beta_1, \beta_2)$ is a solution to the system (3.1). Thus $\beta_1(z,\zeta)$, $\beta_2(z,\zeta)$ and hence $\beta_1{}^*(\zeta,z)$, $\beta_2{}^*(\zeta,z)$ are solutions to (3.1) as are

$$\xi_2 := a\beta_1 + b\beta_2 \quad , \quad \eta_2 := \bar a\beta_1{}^* + \bar b\beta_2{}^*$$

for any choice of the constants a and b. Thus defined $\xi_2{}^* = \eta_2$ and hence it only remains to show that a and b can be chosen so that $J[\xi_2,\eta_2](z^o, \zeta^o) \neq 0$.

We have

$$\begin{aligned} J[\xi_2,\eta_2] &= \partial_{z_2}\xi_2\partial_{\zeta_2}\eta_2 - \partial_{z_2}\eta_2\partial_{\zeta_2}\xi_2 \\ &= a\bar a(\partial_{z_2}\beta_1\partial_{\zeta_2}\beta_1{}^* - \partial_{z_2}\beta_1{}^*\partial_{\zeta_2}\beta_1) \\ &+ \bar a b(\partial_{z_2}\beta_2\partial_{\zeta_2}\beta_1{}^* - \partial_{z_2}\beta_1{}^*\partial_{\zeta_2}\beta_2) \\ &+ a\bar b(\partial_{z_2}\beta_1\partial_{\zeta_2}\beta_2{}^* - \partial_{z_2}\beta_2{}^*\partial_{\zeta_1}\beta_1) \\ &+ b\bar b(\partial_{z_2}\beta_2\partial_{\zeta_2}\beta_2{}^* - \partial_{z_2}\beta_2{}^*\partial_{\zeta_2}\beta_2). \end{aligned}$$

This cannot vanish at (z^o,ζ^o) for all choices of a and b unless all of the parenthetical terms vanish. Subtracting the coefficients of $a\bar a$ and $\bar a b$ and also $a\bar b$ and $b\bar b$ we obtain

$$(\partial_{z_2}\beta_1 - \partial_{z_2}\beta_2)\partial_{\zeta_2}\beta_1{}^* - (\partial_{\zeta_2}\beta_1 - \partial_{\zeta_2}\beta_2)\partial_{z_2}\beta_1{}^* = 0$$

$$(\partial_{z_2}\beta_1 - \partial_{\zeta_2}\beta_2)\partial_{\zeta_2}\beta_2{}^* - (\partial_{\zeta_2}\beta_1 - \partial_{\zeta_2}\beta_2)\partial_{z_2}\beta_2{}^* = 0 .$$

This can happen only if, at (z^o,ζ^o), $\partial_{z_2}\beta_1 = \partial_{z_2}\beta_2$ and $\partial_{\zeta_2}\beta_1 = \partial_{\zeta_2}\beta_2$ which implies $\partial_{z_2}\beta_1\partial_{\zeta_2}\beta_2 - \partial_{z_2}\beta_2\partial_{\zeta_2}\beta_1 = 0$, or if

$$\partial_{\zeta_2}\beta_1{}^*\partial_{z_2}\beta_2{}^* - \partial_{\zeta_2}\beta_2{}^*\partial_{z_2}\beta_1{}^* = (\partial_{z_2}\beta_1\partial_{\zeta_2}\beta_2 - \partial_{z_2}\beta_2\partial_{\zeta_2}\beta_1)^* = 0.$$

Both instances are precluded by (3.3). This completes the proof.

We now proceed with the solution of Case I. For the variables specified in Lemma (3.1) we see that

$$f(z,\zeta) := \tilde{f}(\xi(z,\zeta), \eta(z,\zeta))$$

implies

$$f^*(\zeta,z) = \bar{\tilde{f}}(\xi(\bar{\zeta},\bar{z}), \eta(\bar{\zeta},\bar{z})) = \tilde{f}^*(\eta,\xi).$$

Hence we may take involutions directly with respect to ξ and η. The fixed point about which we seek solutions is now $(\xi^o,\eta^o) := (\xi(z^o,\zeta^o), \eta(z^o,\zeta^o))$. Equation (1.14) can be written

(3.4) $$\partial_{\xi_1} w = - \partial_{\xi_1} \log C^* w .$$

Since $\partial_{\xi_1} w \equiv 0$, this implies immediately the necessary condition

(3.5) $$\partial_{\xi_1\eta_1} \log C^* \equiv 0 .$$

From (3.4)

(3.6) $$w(\xi_1,\xi_2,\eta_2) = w_1(\xi_2,\eta_2)G(\xi_1,\xi_2,\eta_2)$$

where

$$w_1(\xi_2,\eta_2) := w(\xi_1{}^o,\xi_2,\eta_2) , \quad G(\xi_1,\xi_2,\eta_2) := \frac{C^*(\eta_1,\eta_2,\xi_1{}^o,\xi_2)}{C^*(\eta_1,\eta_2,\xi_1,\xi_2)}$$

That $\partial_{\eta_1} G \equiv 0$ follows from (3.5). Substituting (3.6) into (3.4) yields

(3.7) $$\partial_{\eta_2} w_1 + \gamma\partial_{\xi_2} w_1 = Uw_1 + Vw_1{}^*$$

where

$$U := G^{-1}(\partial_{\eta_2} G - \gamma\partial_{\xi_2} G) \qquad V := G^{-1}C(\xi_1,\xi_2,\eta_1{}^o,\eta_2) .$$

We now distinguish between two subcases

<u>Case Ia</u>: $\partial_{\xi_1}\gamma \equiv 0$

<u>Case Ib</u>: There are functions $\tilde{U}_1(\xi_1,\xi_2,\eta_2)$ and $\tilde{V}(\xi_1,\xi_2,\eta_2)$ regular in a

neighborhood of (ξ^o,η^o) such that

(3.8) $$\partial_{\xi_1} U = \tilde{U}\, \partial_{\xi_1}\gamma \qquad\qquad \partial_{\xi_1} V = \tilde{V}\, \partial_{\xi_1}\gamma \ .$$

Theorem 1: *The condition $\partial_{\xi_1}\gamma \equiv 0$ defining Case Ia is equivalent to the condition $B = B(\zeta_1,\zeta_2)$. In this case we may take $\gamma \equiv 0$.*

Proof: By a computation similar to that leading to (1.12) we may establish

(3.9) $$\partial_{\xi_1}\gamma = \partial_{\zeta_2} B^* \frac{J[\xi,\eta](z,\zeta)}{(\partial_{\zeta_2}\eta_2)^2} \ .$$

Hence $\partial_{\zeta_2} B^* \equiv 0$ in Case Ia. The defining condition of Case I is $(\partial_{\zeta_1} + B\partial_{\zeta_2})B^* \equiv 0$ and hence $\partial_{\zeta_1} B^* \equiv 0$ also.

Hence $B := -B_1/B_2$ must be a function of ζ_1 and ζ_2 alone.

In this event we may choose $\xi_2 = f^*(z_1,z_2)$, $\eta_2 = f(\zeta_1,\zeta_2)$ for some analytic f (c.f. Lemma 1). With this choice of ξ_2 and η_2, $\gamma \equiv 0$ and hence the result. Case Ia contains the case studied by Koohara [2].

Let us now proceed with the analysis of Case Ia. Equation (3.7) now has the form

(3.10) $$\partial_{\eta_2} w_1 = Uw_1 + Vw_1^* \ .$$

We make two further distinctions

Case Iai: $\partial_{\xi_1} U = \partial_{\xi_1} V \equiv 0$

Case IIaii: There is a function $H_1(\xi_2,\eta_2)$, regular in a neighborhood of $(\xi_2^{\,o},\eta_2^{\,o})$, such that

(3.11) $$-\,\partial_{\xi_1} U = H_1(\xi_2,\eta_2)\partial_{\xi_1} V \ .$$

Solution to Iai: Since U and V are functions of ξ_2 and η_2 alone (3.10) is a Bers-Vekua equation which may be solved by the method of Volterra integral equations (See [8] and also the solution to Case Ibii). The solution is unique when the analytic function of one variable $w_1(\xi_2,\eta_2^{\,o})$ is specified. The solution to (1.4) is now given by (3.6).

Theorem 2: Suppose $B = B(\zeta_1,\zeta_2)$ $U = U(\xi_2,\eta_2)$, $V = V(\xi_2,\eta_2)$. *Then the system* (1.4) *has a solution depending upon an arbitrary analytic function of one variable.*

Solution to Iaii: We have upon taking the ξ_1-derivative of (3.10)

$$\partial_{\xi_1} Uw_1 + \partial_{\xi_1} Vw_1^* = 0.$$

Hence

(3.12) $$w_1^* = H_1 w .$$

It then follows that for there to be nontrivial solutions

(3.13) $$H_1 H_1^* = 1$$

must hold. From (3.11) there is a regular function H_2 such that

(3.14) $$-U(\xi_1,\xi_2,\eta_2) = H_1(\xi_2,\eta_2)V(\xi_1,\xi_2,\eta_2) + H_2(\xi_2,\eta_2)$$

Equation (3.10) can be written

$$\partial_{\eta_2} w_1 = H_2 w_1 .$$

Hence

(3.15) $$w_1(\xi_2,\eta_2) = w_2(\xi_2) \exp \{ \int_{\eta_2^o}^{\eta_2} H_2(\xi_2,\eta_2{}')d\eta_2{}' \}$$

where $w_2(\xi_2) = w_1(\xi_2,\eta_2{}^o)$. From (3.12) and (3.15)

(3.16) $$w_2^*(\eta_2) = w_2(\xi_2)K(\xi_2,\eta_2)$$

where

$$K(\xi_2,\eta_2) := H_1(\xi_2,\eta_2) \exp \{ \int_{\eta_2^o}^{\eta_2} H_2(\xi_2,\eta_2{}')d\eta_2{}' - \int_{\xi_2^o}^{\xi_2} H_2^*(\eta_2,\xi_2{}')d\xi_2{}' \} .$$

Note

$$KK^* = 1 .$$

Hence

$$w_2(\xi_2) = bK^*(\eta_2{}^o,\xi_2) , \quad w_2^*(\eta_2) = \bar{b}K(\xi_2{}^o,\eta_2) , \quad b \text{ a constant,}$$

and thus from (3.16)

(3.17) $$\frac{\bar{b}K(\xi_2{}^o,\eta_2)K^*(\eta_2,\xi_2)}{K^*(\eta_2{}^o,\xi_2)} = b .$$

Therefore

(3.18) $$|K^*(\eta_2,\xi_2)K^*(\eta_2{}^o,\xi_2)^{-1}K(\xi_2{}^o,\eta_2)| = 1$$

and in view of (3.17) γ depends upon a single arbitrary real constant.

Theorem 3: Suppose $B = B(\zeta_1,\zeta_2)$ and U and V satisfy (3.11). If H_1 and H_2 are such that (3.13) and (3.18) hold, then (1.4) has a solution depending upon an arbitrary real constant.

Let us now consider Case Ib. Taking the ξ_1-derivative of (3.7) we obtain

(3.19) $$\partial_{\xi_2} w_1 = \tilde{U} w_1 + \tilde{V} w_1^* .$$

We again distinguish between two possibilities.

Case Ibi: $\partial_{\xi_1} \tilde{U} = \partial_{\xi_1} \tilde{V} \equiv 0$

Case Ibii: There is a function $L_1(\xi_2,\eta_2)$, regular in a neighborhood of $(\xi_2{}^o,\eta_2{}^o)$ such that

(3.20) $$- \partial_{\xi_1} \tilde{U} = L_1 \partial_{\xi_1} \tilde{V} .$$

Solution to Case Ibi: In this case (3.20) is a Bers-Vekua equation. It may be reduced to the form

(3.21) $$\partial_{\xi_2} w_2 = M w_2^*$$

with

(3.22) $$w_2(\xi_2,\eta_2) = w_1(\xi_2,\eta_2) \exp \{ - \int_{\xi_2^o}^{\xi_2} \tilde{U}(\xi_2{}',\eta_2) d\xi_2{}' \}$$

and

$$M(\eta_2,\xi_2) := \exp \{ \int_{\eta_2^o}^{\eta_2} \tilde{U}^*(\eta_2{}',\xi_2) d\eta_2{}' \; V(\xi_2,\eta_2)$$

If we substitute (3.21) and (3.22) into (3.7) and utilize the fact that $U_1 := U - \gamma\tilde{U}$ and $V_1 = V - \gamma\tilde{V}$ are functions of ξ_2 and η_2 alone, we obtain

(3.23) $$\partial_{\eta_2} w_2 = U_2 w_2 + V_2 w_2^*$$

where

$$U_2(\xi_2,\eta_2) := U_1(\xi_2,\eta_2) + \int_{\xi_2^o}^{\xi_2} \partial_{\eta_2} \tilde{U}(\xi_2{}',\eta_2) d\xi_2{}'$$

$$V_2(\xi_2,\eta_2) := V_1(\xi_2,\eta_2) \exp \{ - \int_{\xi_2^o}^{\xi_2} \tilde{U}(\xi_2{}',\eta_2) d\xi_2{}' + \int_{\eta_2^o}^{\eta_2} \tilde{U}^*(\eta_2{}',\xi_2) d\eta_2{}' \} .$$

Solutions to (3.21) have the representation ([8], p.68)

$$(3.24)\qquad w_2(\xi_2,\eta_2) = \phi(\eta_2) + \int_{\eta_2^o}^{\eta_2} \Gamma_1(\eta_2,\xi_2,\zeta,\xi_2^{\,o})\phi(\tau)d\tau + \int_{\xi_2^o}^{\xi_2} \Gamma_2(\eta_2,\xi_2,\eta_2^{\,o},t)\phi(t)dt$$

where $\phi(\eta_2) = w_2(\xi_2^{\,o},\eta_2)$ and the resolvent kernals Γ_1 and Γ_2 satisfy the relations

$$\Gamma_1(\eta_2,\xi_2,\tau,t) = \int_t^{\xi_2} M^*(\xi_2{}',t)\Gamma_2(\eta_2,\xi_2,t,\xi_2{}')d\xi_2{}'$$

$$(3.25)\qquad \Gamma_2(\eta_2,t,\tau,t) = M(\eta_2,t)$$

$$\Gamma_2(\tau,\xi_2,\tau,t) = M(\tau,t) .$$

From (3.24) and (3.15) we obtain

$$\partial_{\eta_2} w_2(\xi_2^{\,o},\eta_2) = \phi^1(\eta_2)$$

(3.26) and

$$w_2^{\,*}(\eta_2,\xi_2^{\,o}) = \phi^*(\xi_2^{\,o}) + \int_{\eta_2^o}^{\eta_2} M^*(\xi_2^{\,o},t)\phi(t)dt .$$

Thus if we substitute the representation (3.25) with (3.23) and then set $\xi_2 = \xi_2^{\,o}$, we obtain

$$(3.27)\qquad \phi^o(\eta_2) = U_2(\xi_2^{\,o},\eta_2)\phi(\eta_2) + V_2(\xi_2^{\,o},\eta_2)\Big(\phi^*(\xi_2^{\,o}) + \int_{\eta_2^o}^{\eta_2} M^*(\xi_2^{\,o},t)\phi(t)dt\Big) .$$

Let

$$\Phi(\eta_2) := \int_{\eta_2^o}^{\eta_2} M^*(\xi_2^{\,o},t)\phi(t)dt .$$

Then (3.27) becomes

$$(3.28)\qquad \Phi''(\eta_2) + (M^*(\xi_2^{\,o},\eta_2)^{-1}\partial_{\eta_2} M^*(\xi_2^{\,o},\eta_2) - U_2(\xi_2^{\,o},\eta_2))\Phi'(\eta_2) - V_2(\xi_2^{\,o},\eta_2)\Phi(\eta_2) = V_2(\xi_2^{\,o},\eta_2)\phi^*(\xi_2^{\,o}) .$$

The solution of this equation is of the form

$$\Phi = a\psi_1 + b\psi_2 + \phi^*(\xi_2^{\,o})\psi_3$$

where ψ_1 and ψ_2 are solutions to the homogeneous version of equation (3.28) and $\phi^*(\xi_2{}^o)\psi_3$ is a solution to the nonhomogeneous. Thus (3.27) has solution

$$\phi(\eta_2) = M^*(\xi_2{}^o,\eta_2)^{-1}(a\psi_1{}'(\eta_2) + b\psi_2{}'(\eta_2) + \overline{\phi}(\eta_2{}^o)\psi_3{}'(\eta_2)) .$$

One of the constants a and b can be written in terms of $\phi(\eta_2{}^o)$. Hence ϕ has the form

$$\phi = a\tilde{\psi}_1 + \phi(\eta_2{}^o)\tilde{\psi}_2 + \overline{\phi}(\eta_2{}^o)\tilde{\psi}_3$$

where $\tilde{\psi}_1(\eta_2{}^o) = \tilde{\psi}_3(\eta_2{}^o) = 0$ and $\tilde{\psi}_2(\eta_2{}^o) = 1$. From (3.24) w_2 has the form $(c = \phi(\eta_2{}^o))$

$$w_2 = av_1 + \overline{a}v_2 + cv_3 + \overline{c}v_4$$

where the v_j are known functions. Thus equation (3.23) becomes a compatibility condition. The function w_2 depends upon two arbitrary constants providing

$$\text{(3.29)}\qquad \begin{aligned} \partial_{\eta_2} v_1 &= U_2 v_1 + V_2 v_2^* \qquad & \partial_{\eta_2} v_2 &= U_2 v_2 + V_2 v_1^* \\ \partial_{\eta_2} v_3 &= U_2 v_3 + V_2 v_4^* \qquad & \partial_{\eta_2} v_4 &= U_2 v_4 + V_2 v_3^* \end{aligned}$$

are satisfied identically in (ξ_2,η_2). Note that they are satisfied for all points $(\xi_2{}^o,\eta_2)$.

Theorem 4: Suppose $(\partial_{\zeta_1} + B\partial_{\zeta_2})B^* \equiv 0$, and (3.8) is satisfied for some $\tilde{U} = \tilde{U}(\xi_2,\eta_2)$, $\tilde{V} = \tilde{V}(\xi_2,\eta_2)$. Then (1.4) has solutions depending upon two arbitrary complex constants.

Solution to Case Ibii: This case may be solved in a manner similar to that of Iaii. The solutions depend upon one arbitrary real constant.

§4. Solution of Case II

From the defining condition of Case II, (2.4) and (2.6) there are regular functions $D_j(\hat{z}_1,\hat{z}_2,\hat{\zeta}_2)$, $E_j(\hat{z}_1,\hat{z}_2,\hat{\zeta}_2)$, $j = 1,2$ in a neighborhood of $(\hat{z}^o,\hat{\zeta}^o)$ such that

$$\text{(4.1)}\qquad \partial_{z_1}\hat{\zeta}_1 + B^*\partial_{z_2}\hat{\zeta}_2 = D_1 B^* + D_2 ,$$

$$\text{(4.2)}\qquad -(\partial_{\hat{\zeta}_1} \log\hat{C})^* = E_1 B^* + E_2 .$$

Consequently equations (2.2) and (2.5) become

$$\text{(4.3)}\qquad \partial_{\hat{z}_1} w + B^*\partial_{\hat{z}_2} w + (D_1 B^* + D_2)\partial_{\hat{\zeta}_2} w = (E_1 B^* + E_2)w ,$$

$$\text{(4.4)}\qquad \partial_{\hat{z}_2} w + D_1\partial_{\hat{\zeta}_2} w = E_1 w .$$

Let $\tilde{\zeta}_2$ be a solution to the equation

(4.5) $$\partial_{\hat{z}_2}\lambda + D_1\partial_{\hat{\zeta}_2}\lambda = 0$$

for which $\partial_{\hat{\zeta}_2}\tilde{\zeta}_2(\hat{z}^o,\hat{\zeta}^o) \neq 0$. Define a new set of coordinates

$$\tilde{z}_1 = \hat{z}_1 \ , \ \tilde{z}_2 = \hat{z}_2 \ , \ \tilde{\zeta}_1 = \hat{\zeta}_1 \ , \ \tilde{\zeta}_2 = \tilde{\zeta}_2(\hat{z}_1,\hat{z}_2,\hat{\zeta}_2) \ .$$

In these coordinates (4.3) and (4.4) become

(4.6) $$\partial_{\tilde{z}_1} w + B^*\partial_{\tilde{z}_2} w + R\,\partial_{\tilde{\zeta}_2} w = (E_1 B^* + E_2)w$$

(4.7) $$\partial_{\tilde{z}_2} w = E_1 w$$

where $R := \partial_{\hat{z}_1}\tilde{\zeta}_2 + D_2\partial_{\hat{\zeta}_2}\tilde{\zeta}_2$. From 4.7

$$w = w_1(\tilde{z}_1,\tilde{\zeta}_2)\exp\{\int_{\tilde{z}_2^o}^{\tilde{z}_2} E_1(\tilde{z}_1,\tilde{z}_2',\tilde{\zeta}_2)d\tilde{z}_2'\} \ .$$

Substituting this with (4.6) yields

(4.8) $$\partial_{\tilde{z}_1} w_1 + R\,\partial_{\tilde{\zeta}_2} w_1$$

$$= (E_2 - \int_{\tilde{z}_2^o}^{\tilde{z}_2} \partial_{\tilde{z}_1}E_1 d\tilde{z}_2^1 - R\int_{\tilde{z}_2^o}^{\tilde{z}_2} \partial_{\tilde{\zeta}_2} E_1 d\tilde{z}_2'))w_1 \ .$$

Taking the $\tilde{z}_2$-derivative, we obtain

$$\partial_{\tilde{z}_2} R\,\partial_{\tilde{\zeta}_2} w_1$$

$$= \Big[\partial_{\tilde{z}_2} E_2 - \partial_{\tilde{z}_1} E_1 - R\,\partial_{\tilde{z}_2} E_1 - \partial_{\tilde{z}_2} R \int_{\tilde{z}_2^o}^{\tilde{z}_2} \partial_{\tilde{\zeta}_2} E_1 d\tilde{z}_2'\Big]w_1 \ .$$

We consider two subcases.

<u>Case IIa</u>: $\partial_{\tilde{z}_2} R = \partial_{\tilde{z}_2}(\partial_{\hat{\zeta}_1}\tilde{\zeta}_2 + D_2\partial_{\hat{\zeta}_2}\tilde{\zeta}_2) \equiv 0$

<u>Case IIb</u>: There is a function $P_1(\tilde{z}_1,\tilde{\zeta}_2)$ regular in a neighborhood of $(\tilde{z}^o,\tilde{\zeta}^o)$ such that

(4.10) $$\partial_{\tilde{z}_2} E_2 - \partial_{\tilde{z}_1} E_1 - R\,\partial_{\tilde{\zeta}_2} E_1 - \partial_{\tilde{z}_2} R \int_{\tilde{z}_2^o}^{\tilde{z}_2} \partial_{\tilde{\zeta}_2} E_1 d\tilde{z}_2'$$

$$= P_1\partial_{\tilde{z}_2} R \ .$$

Solution to Case IIa: Let us consider the system

$$(\partial_{\zeta_1} + B\partial_{\zeta_2})\eta = 0 .$$

(4.11)

$$(\partial_{z_1} + B^*\partial_{z_2})\eta = 0$$

The first equation is satisfied if $\eta = \eta(\tilde{z}_1,\tilde{z}_2,\tilde{\zeta}_2)$. In terms of these variables the second equation becomes

(4.12) $$\partial_{\tilde{z}_1}\eta + R\,\partial_{\tilde{\zeta}_2}\eta = 0.$$

In view of the defining condition of Case II, (4.12) has a solution $\eta = \eta(\tilde{z}_1,\tilde{\zeta}_2)$ such that $\partial_{\zeta_2}\eta_2(z^o,\zeta^o) \neq 0$. Since η is a solution to (4.11), η^* is also a solution. Let $\eta_2 := a\eta + \overline{a}\eta^*$ where a is chosen so that $\partial_{\zeta_2}\eta_2(z^o,\zeta^o) \neq 0$. If

$$\xi_1 = \tilde{z}_1 ,\ \xi_2 = \tilde{z}_2 ,\ \eta_1 = \tilde{\zeta}_1 ,\ \eta_2 = \eta_2,$$

Then (4.9) can be written

(4.13 $$\partial_{\xi_1} w_1 = (E_2 - \int_{\xi_2^o}^{\xi_2} \partial_{\xi_1} E_1 d\xi_2')w_1 .$$

Moreover $\partial_{\xi_2} w_1 = 0$, and hence we have the necessary condition.

(4.14) $$\partial\xi_2 E_2 - \partial\xi_1 E_1 = 0 .$$

Thus

$$w_1(\xi_1,\eta_2) = w_2(\eta_2)\ \exp \int_{\xi_1^o}^{\xi_1} E_2(\xi_1',\xi_2{}^o,\eta_2)d\xi_1'$$

Substituting into (1.10) we obtain

(4.15) $$w_2'(\eta_2) - \int_{\xi_1^o}^{\xi_1} \partial_{\eta_2} E_2(\xi_1',\xi_2{}^o,\eta_2)d\xi_1'\ w_2(\eta_2)$$

$$+ C\exp\{ - \int_{\xi_1^o}^{\xi_1} E_2(\xi_1',\xi_2{}^o,\eta_2)d\xi_1' + \int_{\eta_1^o}^{\eta_1} E_2{}^*(\eta_1',\overline{\xi}_2{}^o,\eta_2)d\eta_1'\}\ w^*(\eta_2)$$

Necessary conditions are $C = C(\xi_1,\eta_1,\eta_2)$ and

$$\partial_{\eta_1} C \exp \{ \int_{\eta_1^o}^{\eta_1} E_2^*(\eta_1',\bar{\xi}_2^o,\eta_2)d\eta_1' \} = 0 .$$

Hence

$$C = C(\xi_1,\eta_1^o,\eta_2) \exp \{ - \int_{\eta_1^o}^{\eta_1} E_2^*(\eta_1',\bar{\xi}_2^o,\eta_2)d\eta_1' \} \tag{4.16}$$

and thus (4.15) becomes

$$w_2'(\eta_2) = - \int_{\xi_1^o}^{\xi_1} \partial_{\eta_2} E_2(\eta_1',\xi_2^o,\eta_2)d\eta_1' \; w_2(\eta_2) \tag{4.17}$$

$$+ C(\xi_1,\eta_1^o,\eta_2) \exp \{ - \int_{\xi_1^o}^{\xi_1} E_2(\xi_1',\xi_2^o,\eta_2)d\xi_1' \; w_2^*(\eta_2) .$$

We have two subcases

<u>Case IIai</u>: $\partial_{\eta_2} E_2(\xi_1,\xi_2^o,\eta_2) \equiv 0$

<u>Case IIaii</u>: For some $Q_1(\eta_2)$ and $Q_2(\eta_2)$ regular in a neighborhood of (ξ^o,η^o)

$$- \int_{\xi_1^o}^{\xi_1} \partial_{\eta_2} E_{\eta_2}(\xi_1',\xi_2^o,\eta_2)d\xi_1'$$

$$= Q_1(\eta_2)\partial_{\xi_1} [C(\xi_1,\xi_2,\eta_1^o,\eta_2)\exp\{-\int_{\xi_1^o}^{\xi_1} E_2(\xi_1',\xi_2^o,\eta_2)d\xi_1'\}] + Q_2(\eta_2) .$$

<u>Solution to Case IIai</u>: From (4.17)

$$\partial_{\xi_1} [C(\xi_1,\xi_2,\eta_1^o,\eta_2) \exp \{ - \int_{\xi_1^o}^{\xi_1} E_2(\xi_1',\xi_2^o,\eta_2)d\xi_1'\}] = 0$$

and hence

$$C(\xi_1,\xi_2,\eta_1^o,\eta_2) = C(\xi_1^o,\xi_2,\eta_1^o,\eta_2) \exp \{ \int_{\xi_1^o}^{\xi_1} E_2(\xi_1',\xi_2^o,\eta_2)d\xi_1'\} .$$

Thus from (4.12)

(4.18) $$w_2'(\eta_2) = C(\xi_1^o, \eta_1^o, \eta_2) w^*(\eta_2) .$$

This can be solved by a power series argument to yield a solution depending upon one arbitrary complex constant.

Theorem 5: *Suppose* (4.1), (4.2) *and the defining conditions of* Case IIa *and* Case IIai *hold. If* (4.14) *and* (4.16) *are satisfied, then* (1.4) *has solutions depending upon one arbitrary complex constant.*

In Case IIaii, which may be solved in a manner similar to that of Case Iaii the solutions depend upon a single arbitrary real constant.

Solution to Case IIb: From (4.9) we have

$$\partial_{\tilde{\zeta}_2} w_1 = P_1(\tilde{z}_1 \tilde{\zeta}_2) w_1$$

and hence

$$w_1(\tilde{z}_1, \tilde{\zeta}_2) = w_2(\tilde{z}_1) \exp \{ \int_{\tilde{\zeta}_2^o}^{\tilde{\zeta}_2} P_1(\tilde{z}_1, \tilde{\zeta}_2') d\tilde{\zeta}_2' \} .$$

From (4.10) there is a regular function P_2 such that

(4.19) $$E_2 - \int_{\tilde{z}_2^o}^{\tilde{z}_2} \partial_{\tilde{z}_1} E_1 d\tilde{z}_2' - R \int_{\tilde{z}_2^o}^{\tilde{z}_2} \partial_{\tilde{\zeta}_2} E_1 d\tilde{z}_2' = P_1 R + P_2 .$$

Substituting this into (4.8) yields

$$w_2'(\tilde{z}_1) = (P_2 - \int_{\tilde{\zeta}_2^o}^{\tilde{\zeta}_2} \partial_{\tilde{z}_1} P_1) w_2 .$$

A necessary condition is

(4.20) $$\partial_{\tilde{\zeta}_2} P_2 - \partial_{\tilde{z}_1} P_1 = 0 .$$

Hence

$$w_2(\tilde{z}_1) = b \exp \{ \int_{\tilde{z}_1^o}^{\tilde{z}_1} P_2(\tilde{z}_1', \tilde{\zeta}_2^o) d\tilde{z}_1' \} .$$

where b is a constant. Thus

$$w = b \exp\{\int_{\tilde z_1^o}^{\tilde z_1} P_2(\tilde z_1{}', \tilde\zeta_2{}^o)d\tilde z_1' + \int_{\tilde\zeta_2^o}^{\tilde\zeta_2} P_1(\tilde z_1, \tilde\zeta_2{}')d\tilde\zeta_2{}' + \int_{\tilde z_2^o}^{\tilde z_2} E_1(\tilde z_1, \tilde z_2{}', \tilde\zeta_2)d\tilde z_2\}$$

$$=: bT(\tilde z_1, \tilde z_2, \tilde\zeta_2) .$$

From (1.10) is determined up to a real constant by the equation

$$b(P_1(\tilde z_1, \tilde\zeta_2) + \int_{\tilde z_2^o}^{\tilde z_2} \partial_{\tilde\zeta_2} E_1 d\tilde z_2')T(\tilde z_1, \tilde z_2, \tilde\zeta_2)$$

$$= \bar b T^*(\tilde\zeta_1, \tilde\zeta_2, \tilde\zeta_2{}^*(\zeta, z)) .$$

The condition

$$(4.21) \quad |(P_1(\tilde z_1, \tilde\zeta_2) + \int_{\tilde z_2^o}^{\tilde z_2} \partial_{\tilde\zeta_2} E_1 d\tilde z_2)T(\tilde z_1, \tilde z_2, \tilde\zeta_2)T^*(\tilde\zeta_1, \tilde\zeta_2, \tilde\zeta_2{}^*(\zeta, z))^{-1}| = 1$$

is necessary.

Theorem 6: If the defining conditions of Case II and Case IIb and also (4.20) and (4.21) are met, then (1.4) has solutions depending upon an arbitrary real constant.

§5. Solution to Case III.

In this case we may work in the variables $(\hat z, \hat\zeta)$. We may solve successively the equations

$$\partial_{\hat\zeta_2} w = F_1 w ,$$

(2.7) and (2.1). This determines w up to an arbitrary complex constant. Equation (1.8) reduces the latitude to a single real constant. The process and compatibility conditions are similar to those of Case IIb and hence we omit the details.

§6. Conclusion.

We see that in most instances solutions to (1.4) possess little arbitrariness - only in Case Iai do they have even the latitude of an arbitrary analytic function of a single variable. Moreover numerous restraints must be imposed upon the functional forms of the coefficients for there to be solutions at all. Once again only Case Iai has the relatively simple characterization $B_1(z,\zeta) = -B(\zeta)B_z(z,\zeta)$. Nonetheless it is at least possible to find all instances in which there are solutions (excluding situations where the solution process generates singular differential equations) and establish representations for these solutions.

References

[1] Bers, L., Theory of Pseudoanalytic Functions, Institute of Math. and Mech., New York University, New York (1953).

[2] Koohara, A., Similarity principle of the generalized Cauchy-Riemann equations for several complex variables, J. Math. Soc. Japan 23, 213-249 (1971).

[3] Koohara, A., Representation of pseudo-holomorphic functions of several complex variables, J. Math. Soc. Japan 28, 257-277 (1976).

[4] Tutschke, W., Die Cauchysche Integralformel fur morphe Funktionen mehrerer komplexer Variabler, Math. Nachr. 54, 385-391 (1972).

[5] Tutschke, W., Uber die Umwandlung Cauchy nichtlmear partieller Differentialgleichungsysteme, Math. Nachr. 58, 87-136 (1973).

[6] Tutschke,W., Ein Kriterium fur die eindentige Bestmunheit morpher Funkturien mehrer komplexer Variabler, Beitrage zur Analysis 5, 56-62 (1973).

[7] Tutschke, W., A new application of I.N. Vekua's proof to Carlemen's theorem, Soviet Math. Dokl. 15 No. 1, 374-391 (1974).

[8] Vekua, I.N., New Method for Solving Elliptic Equations, Wiley, New York (1967).

[9] Vekua, I.N., Generalized Analytic Functions, Pergamon Press, Oxford (1962).

Contemporary Mathematics
Volume 11, 1982

FOURIER ANALYSIS ON THE UNIT SPHERE : A HYPERCOMPLEX APPROACH

R. Delanghe and F. Sommen (*)

Introduction. Along with the theory of holomorphic functions of several complex variables, which, by the pioneering work of Behnke-Thullen, Cartan and Oka in the thirties, has had an enormous impact on the development of several branches in mathematics and their applications, almost simultaneously other directions have been followed to construct function theories in higher dimension. Hereby we think in particular of quaternionic analysis started by Fueter and Moisil-Theodorescu (see [5] and [14]), which has been used recently to study the Yang-Mills field equations (see [6]). It was also Fueter who introduced the generalized Cauchy-Riemann operator in the framework of Clifford algebra, which in fact extends in a rather natural way both the complex numbers and the real quaternions. As to the use of Clifford algebras in physics we refer to [10,13,18] ; a pure algebraic study of these algebras may be found in [1,17].

In the late sixties the construction of a function theory in this setting was taken up again by Iftimie, Hestenes and Delanghe (see [2,3,11,12]). Iftimie was interested in extending Vekua's theory of generalized analytic functions while Hestenes wished to generalize 3-dimensional vector analytic to n-dimensional space.

The aim of this paper is to show how Clifford analysis may be applied to obtain results in Fourier analysis on the (m+1)-dimensional unit sphere S^m, refining in this way rather recent results (see [8,9,15,16,19]). As we intend to give a survey, no proofs will be worked out. For more details the interested reader is referred to the bibliography given at the end, especially [20], while for a wider information on Clifford analysis and its applications, he is referred to the forthcomming book [4].

One of the main ideas behind Clifford analysis is to develop a theory of functions in Euclidean space which at the same time gener-

---

(*) Research assistant of the National Fund for Sientific Research (Belgium)

0271-4132/82/0000-0143/$04.00

alizes holomorphic function theory of one complex variable and refines the theory of harmonic functions. This is done essentially by linearizing the Laplacian Δ_{m+1} $(m\geq 1)$, or equivalently by factorizing the quadratic form $q(x)=\sum_{i=0}^{m} x_i^2$, $x=(x_0,x_1,\ldots,x_m)\in R^{m+1}$, and this by using a suitable algebra, namely the Clifford algebra A.

If we look at the classical Cauchy-Riemann operator $D=\frac{\partial}{\partial x}+i\frac{\partial}{\partial y}$, then, if $\overline{D}=\frac{\partial}{\partial x}-i\frac{\partial}{\partial y}$, $D\overline{D}=\overline{D}D=\Delta_2=\frac{\partial^2}{\partial x^2}+\frac{\partial^2}{\partial y^2}$.

In the (m+1)-dimensional case, call $D=\sum_{i=0}^{m} e_i\frac{\partial}{\partial x_i}$, where $\{e_1,e_2,\ldots,e_m\}$ $(n\geq m)$ is an orthogonal basis of R^n on which the Clifford algebra A is constructed and where e_0 is the identity of A. Then D can be written as $D=e_0\frac{\partial}{\partial x_0}+(e_1\frac{\partial}{\partial x_1}+e_2\frac{\partial}{\partial x_2}+\ldots+e_m\frac{\partial}{\partial x_m})$ where $e_0\frac{\partial}{\partial x_0}$ and $\sum_{j=1}^{m} e_j\frac{\partial}{\partial x_j}$ are respectively the scalar and vector parts of D. Putting $\overline{D}=e_0\frac{\partial}{\partial x_0}-(\sum_{j=1}^{m} e_j\frac{\partial}{\partial x_j})$, by the multiplication rules $e_ie_j+e_je_i=-2\delta_{ij}e_0$, $i,j=1,\ldots,m$, we get that $D\overline{D}=\overline{D}D=\Delta_{m+1}e_0$.

That Clifford algebras are particularly convenient to construct an algebra-valued function theory was shown in [7].

1. Clifford algebras

1.1. The Clifford algebra was constructed by Clifford in 1878 but not applied to physics until just before 1930. Let us give the construction of a Clifford algebra. We start with an n-dimensional real linear space $V_s^{(n)}$ $(s<n)$ with basis $\{e_1,e_2,\ldots,e_n\}$ and a bilinear form $(v|w)$ such that

$$(e_i|e_j)=0,\ i\neq j,$$
$$(e_i|e_i)=1,\ i=1,\ldots,s,$$
$$(e_i|e_i)=-1,\ i=s+1,\ldots,n.$$

A product is defined in such a way that

(i) $vw+wv=2(v|w)$ for all $v,w\in V_s^{(n)}$;

(ii) the set $C(V_s^{(n)})$ of all real linear combinations of products of vectors from $V_s^{(n)}$ is turned into a linear, associative (but non commutative) algebra over R.

This algebra $C(V_s^{(n)})$ is called a Clifford algebra ; its dimension is 2^n and a basis is given by $\{e_0,e_1,\ldots,e_n,e_1e_2,\ldots,e_{n-1}e_n,\ldots,e_1e_2,\ldots,e_n\}$, where e_0 is the identity.

Notice the following multiplication rules for the basis elements :

$$e_i e_j + e_j e_i = 0,\ i\neq j,\ i=1,\dots,n\ ;$$
$$e_i^2 = 1,\ i=1,\dots,s\ ;$$
$$e_i^2 = -1,\ i=s+1,\dots,n.$$

1.2. If we denote by C_p the subspace spanned by the $\binom{n}{p}$ products $e_{i_1} e_{i_2} \dots e_{i_p}$, then

$$C_+ = \sum_{p\ \text{even}} \oplus\ C_p$$

is a subalgebra of $C(V_s^{(n)})$, called the <u>even subalgebra</u>, which is moreover isomorphic to the Clifford algebra $C(V_{s'}^{(n-1)})$ for some specific $s'\leqslant n$.
As it is well known the Dirac algebra $C(V_1^{(4)})$ has the Pauli algebra $C(V_3^{(3)})$ as its even subalgebra, while in its turn the algebra of real quaternions $C(V_0^{(2)})$ is the even subalgebra of the Pauli algebra. Finally the algebra of complex numbers $C(V_0^{(1)})$ is the even subalgebra of the real quaternions.

1.3. The Clifford algebra which will be considered here is $C(V_0^{(n)})$; it will be denoted by A for short.
An arbitrary basis element of A is given by $e_A = e_{i_1} e_{i_2} \dots e_{i_h}$ where $A=\{i_1,\dots,i_h\}\subset\{1,\dots,n\}$ is ordered in such a way that $1\leqslant i_1<i_2<\dots<i_h\leqslant n$ and where $e_\phi = e_0$ is the identity of A.
Hence any element $a\in A$ may be written as

$$a = \sum_A a_A e_A,\ a_A\in R.$$

If for each e_A we put

$$\overline{e_A} = (-1)^{n_A(n_A+1)/2}\, e_A,$$

where n_A is the cardinality of A, then we call

$$\bar{a} = \sum_A a_A \overline{e_A}.$$

the conjugate of a.
If Re(a) denotes the e_0-coefficient of $a\in A$, then

$$|a|_0^2 = 2^n\ \mathrm{Re}(a\bar{a}) = 2^n \sum_A a_A^2$$

is a norm on A which turns it into a Banach algebra.

2. Clifford analysis

2.1. Let A be the Clifford algebra constructed over $V_o^{(n)}$ and let $\Omega\subset R^{m+1}$ $(1\leq m\leq n)$ be open.

Definition. A function $f\in C_1(\Omega;A)$ is said to be left (resp. right) monogenic in Ω if and only if $Df=0$ (resp. $fD=0$) in Ω. Hereby $D=\sum_{i=0}^{m} e_i \frac{\partial}{\partial x_i}$ and for each $f=\sum_A e_A f_A \in C_1(\Omega;A)$,

$$Df = \sum_{i,A} e_i e_A \frac{\partial f_A}{\partial x_i}, \quad fD = \sum_{i,A} e_A e_i \frac{\partial f_A}{\partial x_i}.$$

In the sequel the space of left (resp. right) monogenic functions in Ω will be denoted by $M_{(r)}(\Omega;A)$ (resp. $M_{(l)}(\Omega;A)$) ; it is a right (resp. left) A-module.

2.2. Cauchy's formula

Put for each $i=0,1,\dots,m$,

$$d\hat{x}_i = dx_0\wedge dx_1\wedge\dots\wedge dx_{i-1}\wedge[dx_i]\wedge dx_{i+1}\wedge\dots\wedge dx_m,$$

$$dx = dx_0\wedge dx_1\wedge\dots\wedge dx_m$$

and

$$d\sigma_x = \sum_{i=0}^{m} (-1)^i \; e_i . d\hat{x}_i .$$

In what follows we shall make use of the function E defined by

$$E(x) = \frac{1}{\omega_{m+1}} \; \frac{\overline{x}}{|x|^{m+1}}$$

where for $x=(x_0,x_1,\dots,x_m)\in R^{m+1}$, $\overline{x} = \sum_{i=0}^{m} x_i\overline{e}_i = x_0 e_0 - \sum_{j=1}^{m} e_j x_j$, $|x|^2 = \sum_{i=0}^{m} x_i^2$ and ω_{m+1} is the area of the unit sphere S^m in R^{m+1}.

Notice that

$$E(x) = \frac{-1}{m-1}\,\overline{D}_x \frac{1}{\omega_{m+1}} \frac{1}{|x|^{m-1}} \quad (m\geq 2)$$

and that it satisfies the following properties :

(i) $E\in C_\infty(R^{m+1}\setminus\{0\};A)$;

(ii) in $R^{m+1}\setminus\{0\}$, $DE = ED=0$;

(iii) $E\in L_1^{loc}(R^{m+1};A)$

and (in the sense of distributions)

(iv) $DE = \delta$;

in other words, E is a fundamental solution of D.
The analogues of all fundamental theorems of classical function theory may be proved, as there are Cauchy's theorem, Liouville's and Morera's theorem, Runge's theorem, Mittag-Leffler's theorem etc. ... We only mention Cauchy's formula.

Theorem (Cauchy's Formula) Let $f \in M_{(r)}(\Omega;A)$ and let S be an (m+1)-dimensional compact differentiable and oriented manifold-with-boundary.
Then for all $x \in \mathring{S}$,

$$f(x) = \int_{\partial S} E(u-x)\,d\sigma_u\, f(u)$$

$$= \frac{1}{\omega_{m+1}} \int_{\partial S} \frac{\overline{u}-\overline{x}}{|u-x|^{m+1}}\, d\sigma_u\, f(u).$$

2.3. A Laurent series

Call for each $l=1,\dots,m$,

$$z_l = x_l e_o - x_o e_l$$

and put for each $(l_1,\dots,l_k) \in \{1,2,\dots,m\}^k$, $k \in \mathbb{N}$,

$$V_{l_1 \dots l_k}(x) = \frac{1}{k!} \sum_{\pi(l_1,\dots,l_k)} z_{l_1} \dots z_{l_k}$$

and

$$W_{l_1 \dots l_k}(x) = (-1)^k\, \partial_{x_{l_1}} \dots \partial_{x_{l_k}} E(x)$$

$$= (-1)^k \frac{1}{\omega_{m+1}}\, \partial_{x_{l_1}} \dots \partial_{x_{l_k}} \frac{\overline{x}}{|x|^{m+1}},$$

where π runs over all distinguishable permutations of the sequence $(l_1,\dots,l_k)$.
Then one may prove that $V_{l_1 \dots l_k}$ and $W_{l_1 \dots l_k}$ are both left and right monogenic in respectively $\mathbb{R}^{m+1}$ and $\mathbb{R}^{m+1}\setminus\{0\}$. In fact they generalize respectively the positive and negative powers of $z=x+iy$.

Definitions. (i) A homogeneous polynomial P_k of degree k which is (left) monogenic in $\mathbb{R}^{m+1}$ is called a (left) inner spherical monogenic of order k.
(ii) A homogeneous function Q_k of order $-(m+k)$ which is (left) monogenic in $\mathbb{R}_o^{m+1} = \mathbb{R}^{m+1}\setminus\{0\}$ is called a (left) outer spherical monogenic of order k.

(iii) The restrictions of the (inner and outer) spherical monogenics to the unit sphere S^m are called surface (inner and outer) spherical monogenics.

Theorem. (Laurent series) Let f be left monogenic in the annular domain $G=\mathring{B}(0,R_2)\setminus\overline{B}(0,R_1)$, $0<R_1<R_2$; then there exists a unique sequence $(P_kf, Q_kf)_{k\in N}$ of inner and outer spherical monogenics such that in G

$$f(x) = \sum_{k=0}^{\infty} P_kf(x) + \sum_{k=0}^{\infty} Q_kf(x),$$

where both series together with all derived series converge absolutely and uniformly on the compact subsets of $\mathring{B}(0,R_2)$ and $R^{m+1}\setminus\overline{B}(0,R_1)$ respectively.
The left inner spherical monogenics P_kf are given by

$$P_kf(x) = \sum_{(l_1,\ldots,l_k)} V_{l_1\ldots l_k}(x)\,\lambda_{l_1\ldots l_k}$$

with

$$\lambda_{l_1\ldots l_k} = \int_{\partial B(0,R)} W_{l_1\ldots l_k}(y)\,d\sigma_y\,f(y),$$

while the left outer spherical monogenics Q_kf are given by

$$Q_kf(x) = \sum_{(l_1,\ldots,l_k)} W_{l_1\ldots l_k}(x)\mu_{l_1\ldots l_k}$$

with

$$\mu_{l_1\ldots l_k} = \int_{\partial B(0,R)} V_{l_1\ldots l_k}(y)\,d\sigma_y\,f(y),$$

$R_1<R<R_2$.

Remarks. (1) (Taylor series) If f is left monogenic in $\mathring{B}(0,R)$, then in $\mathring{B}(0,R)$ f admits the expansion

$$f(x) = \sum_{k=0}^{\infty} P_kf(x)$$

where

$$P_kf(x) = \sum_{(l_1,\ldots,l_k)} V_{l_1\ldots l_k}(x)\,\partial_{x_{l_1}}\ldots\partial_{x_{l_k}} f(0),$$

the series and all derived series being absolutely and uniformly convergent in $\mathring{B}(0,R)$.

(2) Orthogonality relations. We have that

$$\int_{S^m} W_{l_1\ldots l_k}(y)\,d\sigma_y V_{s_1\ldots s_t}(y)$$

$$= \int_{S^m} V_{s_1 \ldots s_t}(y)\, d\sigma_y\, W_{l_1 \ldots l_k}(y)$$

$$= \begin{cases} e_o, & \text{if } (l_1,\ldots,l_k)=(s_1,\ldots,s_t) \\ 0, & \text{otherwise.} \end{cases}$$

3. Fourier analysis on the unit sphere in R^{m+1}

3.1. If one considers the unit circle S^1 in the complex plane, then the set of analytic functions on it, denoted by $\mathcal{A}(S^1)$, may be characterized by

$$\mathcal{A}(S^1) = \lim_{0<\varepsilon<1} \operatorname{ind} \mathcal{O}(\mathring{B}(1+\varepsilon)\setminus \overline{B}(1-\varepsilon))$$

and its dual $\mathcal{A}^*(S^1)$ consists of the so-called analytic functionals in S^1.

Now any $f\in\mathcal{O}(\mathring{B}(1+\varepsilon)\setminus\overline{B}(1-\varepsilon))$ admits a Laurent expansion given by

$$f(z) = \sum_{k=-\infty}^{+\infty} a_k z^k$$

where

$$a_k = \frac{1}{2\pi i} \int_{\partial B(0,\rho)} \frac{f(u)\,du}{u^{k+1}}, \quad 1-\varepsilon<\rho<1+\varepsilon,$$

and so

$$f(z) = \sum_{k=0}^{\infty} \frac{(-1)^k}{k!\,2\pi} \int_{\partial B(0,\rho)} \left[\left(z\frac{d}{du}\right)^k \frac{1}{u}\right] f(u)\frac{du}{i}$$

$$+ \sum_{k=0}^{\infty} \frac{(-1)^k}{k!\,2\pi} \int_{\partial B(0,\rho)} \left[\left(u\frac{d}{dz}\right)^k \frac{1}{z}\right] f(u)\frac{du}{i}.$$

3.2. In the (m+1)-dimensional case, if $f\in M_{(r)}(\mathring{B}(1+\varepsilon)\setminus\overline{B}(1-\varepsilon);A)$, then f admits a Laurent expansion

$$f(x) = \sum_{k=0}^{\infty} P_k f(x) + \sum_{k=0}^{\infty} Q_k f(x)$$

$$= \sum_{k=0}^{\infty} \frac{(-1)^k}{k!\,\omega_{m+1}} \int_{\partial B(0,\rho)} \left[\langle x,\nabla_y\rangle^k \frac{\overline{y}}{|y|^{m+1}}\right] d\sigma_y f(y)$$

$$+ \sum_{k=0}^{\infty} \frac{(-1)^k}{k!\,\omega_{m+1}} \int_{\partial B(0,\rho)} \left[\langle y,\nabla_x\rangle^k \frac{\overline{x}}{|x|^{m+1}}\right] d\sigma_y f(y).$$

So $P_k(x) = |x|^k P_k(\omega)$ and $Q_k(x) = |x|^{-(k+m)} Q_k(\omega)$, $\omega\in S^m$, may be regarded as natural generalizations of $a_k z^k$ and $a_{-(k+1)} z^{-(k+1)}$, $k\in N$.

Theorem. Let f be an analytic A-valued function in S^m, i.e. $f\in\mathcal{A}(S^m;A)$; then f admits a unique expansion in surface spherical monogenics :

$$f(\omega) = \sum_{k=0}^{\infty} P_k f(\omega) + \sum_{k=0}^{\infty} Q_k f(\omega), \ \omega \in S^m.$$

Moreover there exists $0<\varepsilon<1$ such that in $\mathring{B}(1+\varepsilon)\setminus\overline{B}(1-\varepsilon)$ the series $\sum_{k=0}^{\infty} P_k f(x) + \sum_{k=0}^{\infty} Q_k f(x)$ converges to a left monogenic function.

Finally there exists $0<\delta<1$ such that for some $C>0$

$$\sup_{\omega\in S^m} \left\{ \begin{matrix} |P_k f(\omega)|_0 \\ \\ |Q_k f(\omega)|_0 \end{matrix} \right\} \leqslant C(1-\delta)^k.$$

The converse is also true, namely

<u>Theorem</u>. Given a sequence $(P_k(\omega),Q_k(\omega))_{k\in N}$ of surface spherical monogenics satisfying the estimates

$$\sup_{\omega\in S^m} \left\{ \begin{matrix} |P_k(\omega)|_0 \\ |Q_k(\omega)|_0 \end{matrix} \right\} \leqslant C(1-\delta)^k,$$

then

$$f(\omega) = \sum_{k=0}^{\infty} P_k(\omega) + \sum_{k=0}^{\infty} Q_k(\omega)$$

is analytic in S^m.

We may thus conclude that

$$\mathcal{A}(S^m;A) = \lim_{0<\varepsilon<1} \text{ind } M_{(r)}(\mathring{B}(1+\varepsilon)\setminus\overline{B}(1-\varepsilon));A).$$

Of course this characterization is nothing else but a Cauchy-Kowalewski type theorem for functions $f\in (S^m;A)$.

3.3. We also note the following remarkable properties of spherical monogenics.

Consider the (m+1)-dimensional Laplacian

$$\Delta_{m+1} = \sum_{i=0}^{m} \frac{\partial^2}{\partial x_i^2} = \overline{D}D = D\overline{D} \ ;$$

in spherical coordinates $x=r\omega$, $\omega\in S^m$, it becomes

$$\Delta_{m+1} = \frac{\partial^2}{\partial r^2} + \frac{m}{r}\frac{\partial}{\partial r} + \frac{1}{r^2}\Delta^*_{m+1}$$

where Δ^*_{m+1} is the so-called Laplace-Beltrami operator on S^m.

In the same way the generalized Cauchy-Riemann operator $D=\sum_{i=0}^{m} e_i \frac{\partial}{\partial x_i}$ may be written in spherical coordinates $x=r\omega$ as

$$D = \omega\partial_r + \frac{1}{r}\partial_\omega$$

where $\omega = \sum_{i=0}^{m} e_i\omega_i$, $\omega_i = \frac{x_i}{|x|} = \frac{x_i}{r}$,

and

$$\partial_\omega = \sum_{j=1}^{m} \frac{1}{|\frac{\partial\omega}{\partial\theta_j}|^2} \cdot \frac{\partial\omega}{\partial\theta_j} \cdot \frac{\partial}{\partial\theta_j}$$

with

$$|\frac{\partial\omega}{\partial\theta_j}|^2 = \sin^2\theta_1 \ldots \sin^2\theta_{j-1},$$

$(\theta_1,\ldots,\theta_m)$ being the angular coordinates.

If we call $\Gamma = \bar{\omega}\partial_\omega$ the spherical Cauchy-Riemann operator, then we have

Theorem. (i) $\Delta^*_{m+1} = ((m-1)\mathbb{1}-\Gamma)\Gamma$; i.e. a decomposition of the Laplace-Beltrami operator in obtained.

(ii) $\Gamma P_k(\omega) = (-k)P_k(\omega)$ and $\Gamma Q_k(\omega) = (k+m)Q_k(\omega)$; i.e. P_k and Q_k are eigenfunctions of Γ corresponding to the respective eigenvalues $(-k)$ and $(k+m)$.

(iii) If S_k is a spherical harmonic of order k, then $S_k=P_k+Q_{k-1}$; i.e. each spherical harmonic admits a (unique) decomposition into spherical monogenics.

3.4. The decomposition of any spherical harmonic into spherical monogenics gives rise to a number of refinements of results obtained by authors such as Seeley, Hashizume-Kowata-Minemura-Okamoto and Morimoto. It is such that the cited authors have obtained representations of analytic functions, analytic functionals, C_∞-functions, distributions and L_2-functions on the unit sphere S^m in terms of a series of spherical harmonics $\sum_{k=0}^{\infty} S_k(\omega)$, the S_k satisfying some growth conditions depending on the considered case.
We are able to represent each of these functions or functionals f by a series

$$f(\omega) = \sum_{k=0}^{\infty} P_k f(\omega) + \sum_{k=0}^{\infty} Q_k f(\omega),$$

where the sequence $(P_k f, Q_k f)_{k\in\mathbb{N}}$ satisfies similar estimates.
Moreover these f may be represented as boundary values of monogenic

functions in $R^{m+1}\setminus S^m$.

We illustrate these phenomena in the case of $L_2(S^m;A)$, the space of square integrable A-valued functions in S^m.

Definition. For $f,g\in L_2(S^m;A)$ we define the inner product

$$(f,g) = \frac{1}{\omega_{m+1}} \int_{S^m} \overline{f}(\omega)g(\omega)\,dS$$

and the corresponding norm

$$\|f\|^2 = \mathrm{Re}(f,f) = \frac{1}{\omega_{m+1}} \int_{S^m} \mathrm{Re}(\overline{f}(\omega)f(\omega))\,dS,$$

dS being the usual surface measure.

Theorem. Let $f,g\in L_2(S^m;A)$ and let for each $k\in N$,

$$P_k f(x) = \frac{(-1)^k}{k!\,\omega_{m+1}} \int_{S^m} \left[\langle x,\nabla_y\rangle^k \frac{\overline{y}}{|y|^{m+1}} \right]_{y=\omega} d\sigma_\omega f(\omega)$$

and

$$Q_k f(x) = \frac{(-1)^k}{k!\,\omega_{m+1}} \int_{S^m} \left[\langle y,\nabla_x\rangle^k \frac{\overline{x}}{|x|^{m+1}} \right]_{y=\omega} d\sigma_\omega f(\omega).$$

Then

(i) $(f,g) = \sum_{k=0}^{\infty} (P_k f(\omega), P_k g(\omega)) + (Q_k f(\omega), Q_k g(\omega))$;

(ii) $\|f\|^2 = \sum_{k=0}^{\infty} \|P_k f(\omega)\|^2 + \|Q_k f(\omega)\|^2$

(iii) $f(\omega) = \sum_{k=0}^{\infty} P_k f(\omega) + Q_k f(\omega)$, this series being convergent in $L_2(S^m;A)$.

Moreover, if for $f\in L_2(S^m;A)$ we consider its Cauchy transform $\hat{f}$ given by

$$\hat{f}(x) = \int_{S^m} E(x-\omega)\,d\sigma_\omega f(\omega)$$

$$= \sum_{k=0}^{\infty} \chi_{R^{m+1}\setminus\overline{B}(1)} Q_k f(x) - \chi_{\mathring{B}(1)} P_k f(x),$$

where for $F\subset R^{m+1}$, χ_F stands for the characteristic function of F, then we may prove that $\hat{f}\in M_{(r)}(R^{m+1}\setminus S^m)$.

Furthermore

$$\hat{f}(|x|\omega)\in L_2(S^m;A),\quad |x|\neq 1,$$

and

$$\lim_{\varepsilon\to 0+} \hat{f}(\omega(1\pm\varepsilon)) \text{ exists in } L_2(S^m;A).$$

Call for any $f\in L_2(S^m;A)$

$$\Pi_+ f = \lim_{\varepsilon\to 0+} \hat{f}(\omega(1+\varepsilon)) \ ; \ \Pi_- f = \lim_{\varepsilon\to 0+} \hat{f}(\omega(1-\varepsilon)) \ ;$$

then we obtain that Π_+ and Π_- are projections in $L_2(S^m;A)$ satisfying

$$\Pi_+\Pi_- = \Pi_-\Pi_+ = 0$$

and

$$\mathbb{1} = \Pi_+ + \Pi_- .$$

Hence we get the following orthogonal decomposition of $L_2(S^m;A)$ with respect to the (A-valued) inner product (,) defined above :

$$L_2(S^m;A) = \Pi_+ L\,(S^m;A) \oplus_\perp \Pi_- L\,(S^m;A).$$

References

[1] C. Chevalley, The algebraic theory of spinors (Columbia University Press, 1954).

[2] R. Delanghe, On regular-analytic functions with values in a Clifford algebra, Math. Ann. 185 (1970) 91-111.

[3] ——————, On the singularities of functions with values in a Clifford algebra, Math. Ann. 196 (1972) 293-319.

[4] R. Delanghe, F. Brackx, F. Sommen, Clifford analysis, to appear (Research Note Pitman Publishing Limited, London)

[5] R. Fueter, Die Funktionentheorie der Differentialgleichungen $\Delta u=0$ and $\Delta\Delta u=0$ mit vier reellen Variablen, Commentarii Mathematici Helvetici 7 (1934) 307-330.

[6] F. Gürsey, H.C. Tze, Complex and quaternionic analyticity in chiral and gauge theories I, Annals of Physics 128 (1980) 29-130.

[7] K. Habetha, Eine Bemerkung zur Functionentheorie in Algebren, Function theoretic methods for partial differential equations, Proceedings, Lecture Notes in Mathematics 561 (Springer-Verlag, 1976) 502-509

[8] M. Hashizume, A. Kowata, K. Minemura, K. Okamoto, An integral representation of the Laplacian on Euclidean space, Hiroshima Math. J. 2 (1972) 535-545.

[9] S. Helgason, Eigenspaces of the Laplacian ; integral representations and irreducibility, J. Functional Analysis 17 (1974) 328-353.

[10] D. Hestenes, Space-Time algebra (Gordon and Breach, New York, 1966).

[11] ——————, Multivector functions, J. Math. Anal. Appl. 24 (1968) 467-473.

[12] V. Iftimie, Fonctions Hypercomplexes, Bull. Math. Soc. Sci. Math. R.S. Roumanie 9 (57) (1965) 279-332.

[13] P. Lounesto, Spinor valued regular functions in hypercomplex analysis (Ph. D. thesis, Helsinki University of Technology, 1979).

[14] G.C. Moisil, N. Théodorescu, Fonctions holomorphes dans l'espace, Mathematica Cluj 5 (1931) 142-159.

[15] M. Morimoto, Analytic functionals on the sphere and their Fourier-Borel transformations, to appear in a volume of the Banach Center Publication.

[16] ——————, Analytic functionals on the Lie sphere, preprint.

[17] I.R. Porteous, Topological geometry (Van Nostrand, London, 1969)

[18] M. Riesz, Clifford numbers and Spinors, Lecture Series 38 (Institute for Physical Science and Technology, Maryland, 1958).

[19] R.T. Seeley, Eigenfunction expansions of analytic functions, Proc. Amer. Math. Soc. 21 (1969) 734-738.

[20] F. Sommen, Spherical monogenic functions and analytic functionals on the unit sphere, to appear;

Seminar of Algebra and Functional Analysis
State University of Ghent
Krijgslaan 271
B-900 GENT
Belgium

Contemporary Mathematics
Volume 11, 1982

Function Theory for Generalized Beltrami Systems

by

Gerald N. Hile

University of Hawaii

1. INTRODUCTION

The Cauchy-Riemann equations assume the complex form

$$w_x + iw_y = 0 \quad .$$

In [4] A. Douglis developed an analogue of analytic function theory for a more general elliptic system in the plane of the form

$$w_x + iw_y + aEw_x + bEw_y = 0 \quad , \tag{1.1}$$

where E is an mxm constant nilpotent matrix, w is an mx1 vector, and a and b are complex valued functions of x and y. Later in [3] B. Bojarskii extended the function theory of Douglis to a more general system which he wrote in the form

$$w_{\bar{z}} = Q\, w_z \quad . \tag{1.2}$$

He assumed that the variable mxm matrix Q is "lower quasidiagonal" (see section 2) with all the eigenvalues of Q having magnitude less than 1. The systems (1.1) and (1.2) are natural ones to consider because they arise from the reduction of general elliptic systems in the plane to a standard canonical form. This reduction involves changes in both the dependent and independent variables, along with the reduction of the matrix of first order coefficients to Jordan canonical form. This reduction is discussed in both [3] and [4].

Function theory for the Douglis and Bojarskii systems was later further developed by Goldschmidt [10], Hile [11, 12], and Kuhn [15]. In [11] Hile showed that the function theory for elliptic systems with constant coefficients becomes particularly simple because of the simplicity of the generating solution.

This work is an attempt to present results of all these authors and to extend them to a wider class of systems which will be called "generalized Beltrami systems". What appears to be the essential property of elliptic systems in the plane for which one can obtain a useful extension of analytic function theory is the "self-commuting " property of the matrices of first order coefficients. That is to say, the matrix Q in (1.2) satisfies the commuting condition

0271-4132/82/0000-0144/$07.25

$$Q(z_1)\, Q(z_2) = Q(z_2)\, Q(z_1)$$

for any two points z_1, z_2 in the domain of Q. With this condition, plus the ellipticity condition and regularity conditions on Q, we will develop a function theory similar to that of Douglis and Bojarskii. In order to be able later to study exterior boundary value problems, we have also sought to minimize the conditions on Q at infinity. We require that Q be continuous at infinity in such a way that the difference function $Q - Q(\infty)$ is in some $L_p(\mathbb{C})$ class with $1 \leq p < 2$. Under these conditions we prove the existence of an entire generating solution for the system (1.2).

The Douglis and Bojarskii theory has been used to study elliptic systems of the more general form

(1.3) $$w_{\bar{z}} = Qw_z + Aw + B\bar{w} .$$

These results extend the generalized (or "pseudo-") analytic function theory of L. Bers [2] and I. N. Vekua [16]. Work in this direction appears in Begehr and Gilbert [1], Bojarskii [3], Gilbert [6, 7], Gilbert and Hile [8, 9], Goldschmidt [10], and Kuhn [15]. See also the book of Wendland [17] for a more extensive bibliography and a discussion of some of these results. It seems likely that the theory of this paper also may be utilized in similar ways to study the more general system (1.3).

In [13] some aspects of the function theory for elliptic systems has been extended to a Banach algebra setting. The theory is useful in obtaining a generalized notion of the spectrum for elements in a real Banach algebra.

2. GENERALIZED BELTRAMI SYSTEMS AND THE CANONICAL FORM

We consider the first order system of partial differential equations in the plane,

(2.1) $$w_{\bar{z}} = Q\, w_z ,$$

where Q is an mxm complex matrix, and w is mxs and complex. We assume that $Q = Q(z)$ is Holder-continuous in a domain Ω_0 in the complex plane C, by which we mean that for any compact subset K of Ω_0 there exist constants c, α, with $c \geq 0$, $0 < \alpha \leq 1$, such that

$$||Q(z_1) - Q(z_2)|| \leq c|z_1 - z_2|^\alpha \quad , \text{ for all } z_1, z_2 \text{ in K.}$$

We denote this property of Q by writing $Q \in H(\Omega_0)$.

The matrix norm we employ is the standard norm on a matrix $B = (b_{ij})$, given by

$$||B||^2 = \text{trace } (B^*B) = \sum_{i,j} |b_{ij}|^2 \quad .$$

We further require the following commutativity condition on Q,

(2.2) $\quad Q(z_1)\,Q(z_2) = Q(z_2)\,Q(z_1)$, for all z_1, z_2 in Ω_0 .

If Q satisfies (2.2) we say that Q is self-commuting in Ω_0 . More generally, if A and B are matrix valued functions defined in Ω_0 and satisfy the condition

(2.3) $\quad A(z_1)\,B(z_2) = B(z_2)\,A(z_1)$, for all z_1, z_2 in Ω_0 ,

we say that A and B commute in Ω_0 .

Suppose that A and B are of class C^1 and commute in Ω_0 . By considering difference quotients it is easy to verify that all of the derivatives A_x, A_y, A_z, $A_{\bar{z}}$ commute in Ω_0 with B and with all the first derivatives of B. In particular, if A is self-commuting in Ω_0 then the first partial derivatives of A are self-commuting and commute with A and with one another in Ω_0 . We also note that if A commutes with B in Ω_0 then A^{-1} commutes with B in Ω_0 wherever the inverse exists.

As an ellipticity condition on (2.1) we require that Q(z) have no eigenvalues of magnitude 1 at any point z of Ω_0 .

Definition. If Q is self-commuting in Ω_0 , and if Q(z) has no eigenvalues of magnitude 1 for each z in Ω_0 , then system (2.1) is a generalized Beltrami system. Solutions of such a system will be called Q - holomorphic.

It is well known (see [5]) that if Q is Holder-continuous and (2.1) is elliptic then weak solutions w are in fact strong solutions with Holder-continuous first derivatives. This fact will also be a consequence of the canonical form of the system and analogous properties of solutions of a single equation.

Suppose that w and v are both Q - holomorphic in Ω_0 , with w mxm, v mxs, and suppose that w commutes with Q in Ω_0 . Then

$$(wv)_{\bar{z}} = w_{\bar{z}}v + wv_{\bar{z}} = Qw_z v + wQv_z = Q(wv)_z \quad .$$

and therefore the product wv also is Q - holomorphic. Hence if w is mxm, is Q - holomorphic, and commutes with Q in Ω_0, then any positive power w^n is also Q - holomorphic. Moreover, at points where w is invertible we have

$$(w^{-1})_{\bar{z}} = -\,w^{-1}w_{\bar{z}}w^{-1} = -\,w^{-1}Qw_z w^{-1} = -\,Qw^{-1}w_z w^{-1} = Q(w^{-1})_z \quad ,$$

and hence w^{-1} is also Q - holomorphic. We conclude that negative powers of w are Q - holomorphic wherever w is invertible.

A. Douglis in [4] developed a function theory for the system (2.1) where Q has the special form

$$(2.4) \qquad Q = \begin{bmatrix} 0 & & & & \\ a_1 & 0 & & \bigcirc & \\ a_2 & a_1 & 0 & & \\ \vdots & \ddots & \ddots & \ddots & \\ a_{m-1} & \cdots & a_2 & a_1 & 0 \end{bmatrix}$$

He introduced the terminology "generalized Beltrami system" for this special system. Since his function theory extends to our more general system with self-commuting Q, it seems natural also to extend his terminology to this more general system.

In [3] B. Bojarskii extended the function theory of Douglis to the more general case of (2.1) where Q takes the form

$$Q = \begin{bmatrix} A_1 & & & \bigcirc \\ & A_2 & & \\ & & \ddots & \\ \bigcirc & & & A_n \end{bmatrix}$$

and each A_i is a lower triangular block of the form

$$A_i = \begin{bmatrix} \lambda_i & & & & \bigcirc \\ a_{i1} & \lambda_i & & & \\ a_{i2} & a_{i1} & \lambda_i & & \\ \vdots & \ddots & \ddots & \ddots & \\ a_{i,r_i} & \cdots & & a_{i1} & \lambda_i \end{bmatrix}$$

Such matrices Q have the self-commuting property, and were called <u>lower quasidiagonal</u> by Bojarskii. He called solutions of such a system Q - holomorphic. It turns out that the Bojarskii function theory for Q - holomorphic functions carries over with a few modifications to our more general systems. In fact, the canonical form of a generalized Beltrami system is close to the lower quasidiagonal form of Bojarskii. The following lemma is crucial and is a special case of a theorem found in [14], p. 134 :

<u>Lemma 1</u>. Let $\{Q_\alpha\}$ be a commutative collection of complex mxm matrices. Then there exists a nonsingular complex mxm matrix S such that for each Q_α,

$$S\, Q_\alpha\, S^{-1} = \begin{bmatrix} A_{\alpha 1} & & & \bigcirc \\ & A_{\alpha 2} & & \\ & & \ddots & \\ \bigcirc & & & A_{\alpha n} \end{bmatrix}$$

and each $A_{\alpha i}$ has the form

$$A_{\alpha i} = \begin{bmatrix} \rho_{\alpha i} & & & 0 \\ & \rho_{\alpha i} & & \\ & & \ddots & \\ * & & & \rho_{\alpha i} \end{bmatrix} .$$

(The size of each square block $A_{\alpha i}$ is the same for all α, but may change with i. The * denotes possibly nonzero terms.)

The preceding lemma shows that the self-commuting condition is indeed very strong. We have the following consequence for the system (2.1) :

Theorem 2. Let (2.1) be a generalized Beltrami system in Ω_0. Then there exists a constant complex mxm matrix S such that if $v = Sw$ and $\hat{Q} = SQS^{-1}$, then (2.1) transforms into the equation

$$v_{\bar{z}} = \hat{Q}\, v_z \qquad (2.5)$$

Moreover, $\hat{Q}$ has the form

$$\hat{Q}(z) = \begin{bmatrix} A_1(z) & & & 0 \\ & A_2(z) & & \\ & & \ddots & \\ 0 & & & A_n(z) \end{bmatrix} , \qquad (2.6)$$

where each A_i is a block of the form

$$A_i(z) = \begin{bmatrix} \lambda_i(z) & & & 0 \\ & \lambda_i(z) & & \\ & & \ddots & \\ * & & & \lambda_i(z) \end{bmatrix} , \qquad (2.7)$$

and the size of each block $A_i(z)$ is the same for all z.

Proof. Equation (2.5) is easily verified. As S we take the constant matrix S guaranteed by Lemma 1 applied to the collection $\{Q(z)\}$, $z \in \Omega_0$.

We note that since $\hat{Q}$ is lower triangular, the $\lambda_i(z)$ in (2.7) are the eigenvalues of $\hat{Q}(z)$. Since the eigenvalues are invariant under a similarity transformation, the $\lambda_i(z)$ are also the eigenvalues of $Q(z)$. We also see that $\hat{Q}$ and the blocks A_i have the self-commuting property in Ω_0. Each $A_i(z)$ may be written in the form

$$A_i(z) = \lambda_i(z)\, I + N_i(z)$$

where I is the identity matrix and $N_i(z)$ is a lower triangular nilpotent matrix with zeros on the main diagonal. One sees that each $N_i(z)$ also is self-commuting in Ω_0 .

From (2.6) and (2.7) we see that the canonical form (2.5) of system (2.1) separates into n disjoint subsystems of the form

(2.8)
$$v_{\bar{z}} = A\, v_z \quad ,$$

where A is a matrix of the form (2.7), and the diagonal term, $\lambda(z)$, satisfies $|\lambda(z)| \neq 1$ in Ω_0 . The first equation in (2.8) involving only the first component v_1 of v is an ordinary Beltrami equation,

$$v_{1,\bar{z}} = \lambda v_{1,z} \quad .$$

Since λ is Holder-continuous, any solution v_1 of this equation has Holder-continuous first derivatives. The second equation of the system (2.8) is of the form

$$v_{2,\bar{z}} = \lambda v_{2,z} + a v_{1,z} \quad .$$

This equation is a nonhomogeneous Beltrami equation in v_2 with Holder-continuous coefficients, and it follows therefore that v_2 has Holder-continuous first derivatives. Proceeding in this manner, one can show by using properties of solutions of ordinary nonhomogeneous Beltrami equations that solutions of (2.5), and hence of the generalized Beltrami system (2.1), have Holder-continuous partial derivatives of the first order.

An obvious question is whether the matrix $\hat{Q}$ of (2.6) can be brought into the quasidiagonal form of Bojarskii by a similarity transformation, which would imply that the Bojarskii form is sufficiently general to serve as a canonical form for all generalized Beltrami systems. We answer this question in the negative by presenting as a counterexample the 3x3 matrix

$$Q(z) = \begin{bmatrix} i & 0 & 0 \\ q(z) & i & 0 \\ r(z) & 0 & i \end{bmatrix} \quad .$$

This matrix Q is self-commuting for any choices of the functions q(z) and r(z). For the choices $q(z) = 1$, $r(z) = x$, it is easy to see that there is no invertible constant matrix S such that $SQ = \hat{Q}S$ and such that $\hat{Q}$ has the Bojarskii form. Even if we allow S to be a variable but continuous matrix there are counterexamples. For example, if

$$q(z) = x^n \sin(1/x) + x^{n+1}\cos(1/x) \quad , \quad r(z) = x^{n+1}\sin(1/x) + x^n \cos(1/x) \quad ,$$

where n is a positive integer, then Q is of class C^{n-1}. One can show that in any neighborhood of any point on the y-axis there is no invertible and continuous matrix $S = S(z)$ such that $SQ = \hat{Q}S$ and such that $\hat{Q}$ has the Bojarskii form. (The proof of this last statement is somewhat detailed and will not be presented here.)

3. THE P MATRIX FOR A GENERALIZED BELTRAMI SYSTEM

Let Q be a constant mxm complex matrix with no eigenvalues of magnitude one. (Later we allow Q to be a function of z.) We wish to evaluate the integral

$$P = \int_{|z|=1} (z\,I + \bar{z}\,Q)^{-1}\,(I\,dz + Q\,d\bar{z}) \quad . \tag{3.1}$$

If $Q = 0$ then $P = 2\pi i I$. For Q - holomorphic functions P will play the same role as the constant $2\pi i$ plays for analytic functions. Since no eigenvalues of Q have magnitude one the matrix $(z\,I + \bar{z}\,Q)$ is invertible on $|z| = 1$ and therefore the integral (3.1) is defined.

Theorem 3. The matrix P, obtained from Q by formula (3.1), is invertible and commutes with Q, and the only possible eigenvalues of P are $\pm\, 2\pi i$. If all the eigenvalues of Q are of magnitude less than one, then $P = 2\pi i I$, and if all the eigenvalues of Q have magnitude greater than one, then $P = -\, 2\pi i I$.

Proof. We prove the theorem in stages. First assume that $Q = \lambda I$, where $|\lambda| < 1$. Then

$$P = I \int_{|z|=1} \frac{1}{z + \lambda\bar{z}}\,(dz + \lambda d\bar{z}\,) \quad . \tag{3.2}$$

Since $|\lambda| < 1$, the expression $z + \lambda\bar{z}$ never vanishes on $|z| = 1$, and as z traces a path around the unit circle in the positive direction the change in the argument of $z + \lambda\bar{z}$ is the same as the change in the argument of z, namely 2π . Thus, for an appropriate branch of the complex logarithm function we have

$$P = I \int_{|z|=1} d\,[\log\,(z + \lambda\bar{z})] = 2\pi i I \quad .$$

On the other hand, if $Q = \lambda I$, where $|\lambda| > 1$, we make the change of variables $\zeta = \bar{z}$ in (3.2) to obtain

$$P = -I \int_{|\zeta|=1} (\bar{\zeta} + \lambda\zeta)^{-1}\,(d\bar{\zeta} + \lambda d\zeta\,) = -I \int_{|\zeta|=1} (\zeta + \lambda^{-1}\bar{\zeta}\,)^{-1}\,(d\zeta + \lambda^{-1} d\bar{\zeta}\,)$$

$$= -\,2\pi i I \quad (\text{since } |\lambda^{-1}| < 1\,) .$$

Next assume that Q has the form $Q = \lambda I + N$, where $|\lambda| \neq 1$ and N is a constant nilpotent matrix satisfying $N^r = 0$ for some positive integer r. Then

$$(zI + \bar{z}Q)^{-1} = [(z + \lambda\bar{z})I + \bar{z}N]^{-1} = \sum_{k=0}^{r-1} (-1)^k (z + \lambda\bar{z})^{-k-1}(\bar{z})^k N^k$$

and

$$(zI + \bar{z}\,Q)^{-1} (I\,dz + Q\,d\bar{z}) = (zI + \bar{z}\,Q)^{-1} (I\,dz + \lambda I\,d\bar{z} + N\,d\bar{z})$$

$$= (z + \lambda\bar{z})^{-1} (dz + \lambda\,d\bar{z})\,I - \sum_{k=1}^{r-1} (-1)^k k^{-1} d[(z + \lambda\bar{z})^{-k} (\bar{z})^k N^k] \quad .$$

If this expression is integrated around the circle $|z| = 1$ the terms involving a total differential vanish, and we obtain

$$P = I \int_{|z|=1} (z + \lambda\bar{z})^{-1} (dz + \lambda\,d\bar{z}) = \pm 2\pi i\,I \quad ,$$

where the plus sign is chosen if $|\lambda| < 1$ and the minus sign if $|\lambda| > 1$.

Next assume that Q has the form of $\hat{Q}$ in (2.6). Since each A_i has the form $A_i = \lambda_i I + N_i$, where N_i is nilpotent, the corresponding P value, $\hat{P}$, for $\hat{Q}$ is

$$(3.3) \qquad \hat{P} = \begin{bmatrix} \pm 2\pi i I_1 & & & 0 \\ & \pm 2\pi i I_2 & & \\ & & \ddots & \\ 0 & & & \pm 2\pi i I_n \end{bmatrix}$$

where each I_i is the identity matrix of the same size as A_i , and the plus or minus signs are chosen according to whether λ_i lies inside or outside the unit disk. If all the eigenvalues are inside (or outside) the unit disk, then $\hat{P} = 2\pi i I$ (or $-2\pi i I$).

Finally, if Q is an arbitrary matrix with no eigenvalues of magnitude one, let $\hat{Q} = SQS^{-1}$, where $\hat{Q}$ has the form (2.6). Let $\hat{P}$ be the corresponding P value for $\hat{Q}$. Then

$$SPS^{-1} = \int_{|z|=1} S(zI + \bar{z}\,Q)^{-1} S^{-1} S(I\,dz + Q\,d\bar{z})\,S^{-1}$$

$$= \int_{|z|=1} (zI + \bar{z}\,\hat{Q})\,(I\,dz + \hat{Q}\,d\bar{z}) = \hat{P} \quad ,$$

and thus $P = S^{-1}\hat{P}S$. Therefore the eigenvalues of P are the same as those of $\hat{P}$, namely $\pm 2\pi i$. If all the eigenvalues of Q lie inside (or outside) the unit disk, then $P = \hat{P} = 2\pi i I$ (or $-2\pi i I$).

The statement that P commutes with Q is clear from the integral representation (3.1).

Now we allow the case where $Q = Q(z)$ is a variable and self-commuting matrix defined in a domain Ω_0. Then the associated matrix $P = P(z)$ formally depends on the point z. However, surprisingly, it turns out that if Q is continuous then $P(z)$ is the same for all points z in Ω_0.

Theorem 4. Let $Q = Q(z)$ be an mxm complex matrix defined and continuous in a domain Ω_0 in the plane. Suppose that Q is self-commuting in Ω_0, and, for each z in Ω_0, $Q(z)$ has no eigenvalues of magnitude one. Let

$$(3.4) \qquad P(z) = \int_{|\zeta|=1} [\zeta I + \bar{\zeta} Q(z)]^{-1} (I\, d\zeta + Q(z)\, d\bar{\zeta}) \quad .$$

Then $P(z)$ is constant in Ω_0, and moreover commutes with Q in Ω_0.

Proof. By Theorem 2 there exists a nonsingular matrix S such that for each z in Ω_0 the matrix $\hat{Q}(z) = SQ(z)S^{-1}$ has the form (2.6). The corresponding $\hat{P}(z)$ associated with $\hat{Q}(z)$ has the form (3.3). Since Q and hence $\hat{Q}$ are continuous functions of z it follows from the representation (3.4) for $\hat{P}$ that $\hat{P}$ also is a continuous function of z. Thus the choice for the $\pm$ sign on each block in (3.3) is the same for all z in Ω_0, and $\hat{P}$ is a constant matrix. Hence $P = S^{-1}\hat{P}S$ is also constant.

Again, the fact that P commutes with Q in Ω_0 follows from the integral representation (3.4) for P.

The constant matrix P defined by (3.4) will be called the P value for the generalized Beltrami system (2.1). If for each z all the eigenvalues of $Q(z)$ lie inside (or outside) the unit disk, then $P = 2\pi i I$ (or $-2\pi i I$). More generally, we see from the proofs of Theorems 3 and 4 that if the sum of the multiplicities of the eigenvalues of Q inside the unit disk is r, and the sum of the multiplicities of the eigenvalues outside the unit disk is s, then for P the multiplicity of the eigenvalue $2\pi i$ is r and the multiplicity of the eigenvalue $-2\pi i$ is s.

4. GENERATING SOLUTIONS

Following Douglis and Bojarskii, we introduce the concept of a generating solution for the generalized Beltrami system

$$(4.1) \qquad w_{\bar{z}} = Q\, w_z \quad .$$

Definition. A generating solution for (4.1) in a domain Ω_0 is an mxm complex matrix valued function $\phi = \phi(z)$ defined in Ω_0 and having the following properties :

(G1) ϕ is a C^1 solution of (4.1) in Ω_0 .

(G2) ϕ is self-commuting in Ω_0 and commutes with Q in Ω_0 .

(G3) $\phi(\zeta) - \phi(z)$ is invertible for all z, ζ in Ω_0 , $z \neq \zeta$.

(G4) $\phi_z(z)$ is invertible for all z in Ω_0 .

We note that these conditions do not necessarily determine the generating solution uniquely. For example, if (4.1) is the single complex Cauchy-Riemann equation $w_{\bar{z}} = 0$, then the functions $\phi_1(z) = z$ and $\phi_2(z) = \sin z$ are both generating solutions inside the strip $|\text{Re } z| < 1$.

For the case when Q is a constant matrix it is easily checked that a generating solution is

$$\phi(z) = zI + \bar{z}\,Q \qquad .$$

The theory in this case is greatly simplified, as will be pointed out in section 6.

In section 7 we will prove existence of a generating solution under rather general conditions on Q. For this section and the next we assume that a generating solution exists in Ω_0 and develop consequences of this fact.

Since we are assuming that Q is in $H(\Omega_0)$ the first derivatives of ϕ are also in $H(\Omega_0)$. Let $z, \zeta \in \Omega_0$, and suppose that the straight line connecting z and ζ lies in Ω_0 . Then

$$\phi(\zeta) - \phi(z) - \phi_z(z)(\zeta - z) - \phi_{\bar{z}}(z)(\bar{\zeta} - \bar{z})$$

$$= \int_0^1 \{ [\phi_z(z + t(\zeta-z)) - \phi_z(z)](\zeta-z) + [\phi_{\bar{z}}(z + t(\zeta-z)) - \phi_{\bar{z}}(z)](\bar{\zeta}-\bar{z}) \}\, dt$$

Since ϕ_z and $\phi_{\bar{z}}$ are in $H(\Omega_0)$, there exist constants c, α, with $c \geq 0$, $0 < \alpha \leq 1$, such that for all ζ sufficiently close to z,

$$||\phi(\zeta) - \phi(z) - \phi_z(z)(\zeta-z) - \phi_{\bar{z}}(z)(\bar{\zeta}-\bar{z})|| \leq \int_0^1 ct^{\alpha}\, |\zeta - z|^{\alpha+1}\, dt$$

$$= \frac{c}{\alpha + 1}\, |\zeta - z|^{\alpha+1} \qquad .$$

Thus, since ϕ also satisfies (4.1), the following property holds :

(G5) $\phi(\zeta) - \phi(z) = [(\zeta - z)\, I + (\bar{\zeta} - \bar{z}\,)\, Q(z)]\, \phi_z(z) \; + \; O(|\zeta - z|^{\alpha+1}\,)$,

as $\zeta \to z$, $z \in \Omega_0$.

(We note that α depends on z, but may be regarded as constant as z ranges over any compact subset of Ω_0 .)

From (G5) we obtain also the estimate

$$[\phi(\zeta) - \phi(z)]^{-1} = (\overline{\zeta}-\overline{z})^{-1}[\frac{\zeta-z}{\overline{\zeta}-\overline{z}} I + Q(z)]^{-1}\{\phi_z(z) + [\frac{\zeta-z}{\overline{\zeta}-\overline{z}}I + Q(z)]^{-1}\mathcal{O}(|\zeta-z|^{\alpha})\}^{-1} .$$

Since Q(z) has no eigenvalues of magnitude one, the matrices

$$[\frac{\zeta - z}{\overline{\zeta} - \overline{z}} I + Q(z)]^{-1}$$

are uniformly bounded for all $\zeta \neq z$. Thus we obtain the additional estimate for the generating solution,

$$\text{(G6)} \qquad ||[\phi(\zeta) - \phi(z)]^{-1}|| = \mathcal{O}(|\zeta - z|^{-1}) \quad \text{as } \zeta \to z , z \in \Omega_0 .$$

The following result was proved by Goldschmidt [10] under slightly stronger conditions on Q. Here we give a somewhat more elementary proof.

Theorem 5. Let w = w(z) be an mxs complex matrix valued function in $C^1(\Omega_0)$. Then the limit

$$\text{(4.2)} \qquad \frac{dw(z)}{d\phi} \equiv \lim_{\Delta z \to 0} [\phi(z + \Delta z) - \phi(z)]^{-1}[w(z + \Delta z) - w(z)]$$

exists at a point z in Ω_0 if and only if w satisfies (4.1) at z, and in this case,

$$\text{(4.3)} \qquad \frac{dw(z)}{d\phi} = \phi_z(z)^{-1} w_z(z) .$$

Proof. First suppose that the limit (4.2) exists. Property (G5) implies

$$\phi(z + \Delta z) - \phi(z) = [(\Delta z)I + (\Delta\overline{z})Q(z)] \phi_z(z) + \mathcal{O}(|\Delta z|^{\alpha+1})$$

$$= \{[I + \frac{\Delta\overline{z}}{\Delta z} Q(z)] \phi_z(z) + \mathcal{O}(|\Delta z|^{\alpha})\} \Delta z .$$

Therefore, setting $\Delta z = \Delta x$ in (4.2) and letting $\Delta z \to 0$, we obtain

$$\text{(4.4)} \qquad \frac{dw(z)}{d\phi} = [I + Q(z)]^{-1}\phi_z(z)^{-1} w_x(z) .$$

Similarly, setting $\Delta z = i\Delta y$, we obtain

$$\text{(4.5)} \qquad \frac{dw(z)}{d\phi} = [I - Q(z)]^{-1} \phi_z(z)^{-1} (-iw_y(z)) .$$

Equating (4.4) and (4.5) we obtain

$$(I + Q)^{-1} w_x = (I - Q)^{-1} (-iw_y) ,$$

which yields (4.1). Now, conversely, we assume that (4.1) holds at z, and show that the limit (4.2) exists. Property (G5) for ϕ may be derived in exactly the same way for w. Thus we have, for some α,

$$(4.6)\quad [\phi(z+\Delta z)-\phi(z)]^{-1}\,[w(z+\Delta z)-w(z)]$$

$$=\{\phi_z(z)+[\frac{\Delta z}{\Delta\bar z}I+Q(z)]^{-1}O(|\Delta z|^{\alpha})\}^{-1}\{w_z(z)+[\frac{\Delta z}{\Delta\bar z}I+Q(z)]^{-1}O(|\Delta z|^{\alpha})\}$$

Since no eigenvalues of Q have magnitude one, and $|(\Delta z)/(\Delta\bar z)| = 1$, the matrices

$$[\ \frac{\Delta z}{\Delta\bar z}I+Q(z)\ ]^{-1}$$

are uniformly bounded for all $\Delta z \neq 0$. Thus, letting Δz approach zero in (4.6), we obtain (4.3).

We adopt Goldschmidt's terminology and say that w is ϕ - differentiable if w is of class C^1 and the limit (4.2) exists. It is clear that any integral power ϕ^n, $n \geq 0$, is ϕ - differentiable in Ω_o , and

$$\frac{d}{d\phi}(\phi^n) = n\phi^{n-1}\quad .$$

Moreover, at all points in Ω_o except $z = \zeta$ we have for fixed ζ,

$$\frac{d}{d\phi}[\phi(\zeta)-\phi]^{-n} = n[\phi(\zeta)-\phi]^{-n-1}\quad .$$

5. FUNCTION THEORY FOR A GENERALIZED BELTRAMI SYSTEM

Following the general methods of Douglis [4] and Bojarskii [3] we develop a function theory for the generalized Beltrami system,

$$(5.1)\qquad w_{\bar z} = Q\,w_z\quad .$$

First we derive a Green's formula associated with equation (5.1). Similar Green's formulas were used by Douglis and Bojarskii under slightly stronger conditions Q.

We say that a bounded domain Ω is a regular subdomain of Ω_o if $\bar\Omega \subset \Omega_o$, and if the boundary Γ of Ω consists of a finite number of simple closed curves with piecewise continuous tangent.

If v and u are mxm and mxs matrices, respectively, we formally define the product

$$(dv)u = v_z\,u\,dz + v_{\bar z}\,u\,d\bar z\quad .$$

Since matrices of course need not commute, the order of multiplications must be observed.

Theorem 6. Let v be an mxm solution of (5.1) in Ω_0 which is in $C^1(\Omega_0)$ and commutes with Q in Ω_0. Let Ω be a regular subdomain of Ω_0, with $\Gamma = \partial\Omega$. If $u \in C^1(\Omega) \cap C(\overline{\Omega})$ and has bounded first derivatives in Ω, then

$$(5.2) \qquad \int_\Gamma (dv)u = 2i \iint_\Omega v_z (u_{\overline{z}} - Q u_z) \, dx\, dy$$

Proof. First assume that v is in $C^\infty(\mathbb{C})$. Then by the ordinary complex Green's formula,

$$(5.3) \qquad \begin{aligned} \int_\Gamma (dv)\, u &= \int_\Gamma (v_z u \, dz + v_{\overline{z}} u \, d\overline{z}) \\ &= 2i \iint_\Omega [(v_z u)_{\overline{z}} - (v_{\overline{z}} u)_z] \, dx\, dy = 2i \iint_\Omega (v_z u_{\overline{z}} - v_{\overline{z}} u_z) \, dx\, dy . \end{aligned}$$

The first and last integrals of (5.3) involve only the first derivatives of v. Thus if v is only of class $C^1(\Omega_0)$ we may apply (5.3) to a sequence of functions in $C^\infty(\mathbb{C})$ which approximate v in the $C^1(\overline{\Omega})$ norm to conclude that the first and last integrals of (5.3) are equal if v is only of class $C^1(\Omega_0)$. We then obtain (5.2) by setting $v_{\overline{z}} = Qv_z$ in (5.3) and using the commutativity of v_z and Q.

The following corollary is immediate.

Corollary 7. Let ϕ be a generating solution for (5.1) in Ω_0, and let Ω be a regular subdomain of Ω_0 with boundary Γ. If w is in $C(\overline{\Omega})$ and Q - holomorphic in Ω, then

$$\int_\Gamma (d\phi)\, w = 0 .$$

The next lemma is preliminary to the derivation of a Cauchy integral formula.

Lemma 8. Let ϕ be a generating solution for (5.1) in Ω_0. If $z \in \Omega_0$ and if the closed disk of radius ε about z is also contained in Ω_0, then

$$(5.4) \qquad \int_{|\zeta - z| = \varepsilon} [\phi(\zeta) - \phi(z)]^{-1} \, (d\phi)(\zeta) = P ,$$

where P is the P - matrix for the system (5.1).

Proof. Corollary 7 implies that the value of the integral (5.4) is independent of ε, and in fact the value of this integral does not change if the curve $|\zeta - z| = \varepsilon$ is deformed into any smooth curve about z with winding number one. Thus, to compute the value of (5.4) it is sufficient to find the limit of the value of this integral as ε tends to zero. We use properties (G1) and (G5) for ϕ and set $\zeta - z = \varepsilon e^{i\theta}$ to obtain

$$[\phi(\zeta) - \phi(z)]^{-1}\, d\phi(\zeta)$$

$$= \{[(\zeta - z)I + (\overline{\zeta} - \overline{z})Q(z)]\, \phi_z(z) + O(|\zeta-z|^{1+\alpha})\}^{-1}\phi_\zeta(\zeta)[d\zeta + Q(\zeta)\, d\overline{\zeta}\,]$$

$$= \{[e^{i\theta}I+e^{-i\theta}Q(z)]\phi_z(z) + O(\varepsilon^\alpha)\}^{-1}\phi_\zeta(z+\varepsilon e^{i\theta})[ie^{i\theta}-Q(z+\varepsilon e^{i\theta})ie^{-i\theta}\,]\, d\theta \quad .$$

Substituting this expression into the left side of (5.4) and letting ε tend to zero we obtain as the limit of the left side the value

$$\int_0^{2\pi} [\, e^{i\theta}\, I + e^{-i\theta}\, Q(z)]\, [Ie^{i\theta} - Q(z)ie^{-i\theta}\,]\, d\theta$$

$$= \int_{|\zeta|=1} [\, \zeta I + \overline{\zeta}Q(z)\,]^{-1}\, [I\, d\zeta + Q(z)\, d\overline{\zeta}\,] \quad = \quad P \qquad .$$

For the remainder of this section we let ϕ denote a given and fixed generating solution for (5.1) in a domain Ω_0 .

Theorem 9. Let Ω be a regular subdomain of Ω_0 , with $\Gamma = \partial\Omega$, and let w be an mxs matrix in $C^1(\Omega) \cap C(\overline{\Omega})$ with bounded first derivatives in Ω. Then for z in Ω,

(5.5) $$w(z) = P^{-1}\int_\Gamma [\, \phi(\zeta) - \phi(z)\,]^{-1}\, (d\phi)(\zeta)\, w(\zeta)$$

$$- 2iP^{-1}\iint_\Omega \phi_\zeta(\zeta)\, [\phi(\zeta) - \phi(z)]^{-1}[w_{\overline{\zeta}}(\zeta) - Q(\zeta)w_\zeta(\zeta)]\, d\xi\, d\eta$$

(where $\zeta = \xi + i\eta$) .

Proof. Fix z in Ω, and let Ω_ε be the domain Ω with a small disk of radius ε about z deleted. Apply (5.2) with $v(\zeta) \equiv \phi(\zeta)$, $u(\zeta) \equiv [\phi(\zeta) - \phi(z)]^{-1}w(\zeta)$, to obtain

$$2i\iint_{\Omega_\varepsilon} \phi_\zeta(\zeta)\, [\, \phi(\zeta) - \phi(z)\,]^{-1}\, [\, w_{\overline{\zeta}}(\zeta) - Q(\zeta)\, w_\zeta(\zeta)\,]\, d\xi\, d\eta$$

$$= \int_\Gamma (d\phi)(\zeta)[\phi(\zeta) - \phi(z)]^{-1}w(\zeta)\, d\xi\, d\eta$$

(5.6) $$- \int_{|\zeta-z|=\varepsilon} (d\phi)(\zeta)\, [\phi(\zeta) - \phi(z)]^{-1}\, w(\zeta)\, d\xi\, d\eta \qquad .$$

We wish to evaluate the limit of the last integral as $\varepsilon \to 0$. Applying (5.4) and (G6) we obtain the estimates

$$\|\int_{|\zeta-z|=\varepsilon} d\phi(\zeta)\, [\phi(\zeta) - \phi(z)\,]^{-1}w(\zeta)\, d\xi\, d\eta \; - \; Pw(z)\, \|$$

$$\leq \int_{|\zeta-z|=\varepsilon} \|\, [\phi(\zeta) - \phi(z)\,]^{-1}\|\; \|\, w(\zeta) - w(z)\, \|\; \|d\phi(\zeta)\|$$

$$\leq (\text{constant}) \int_{|\zeta-z|=\varepsilon} |\zeta-z|^{-1} \, ||w(\zeta)-w(z)|| \, ||\phi_\zeta(\zeta)\, d\zeta + \phi_{\overline{\zeta}}(\zeta)\, d\overline{\zeta}\, ||$$

$$\leq (\text{constant}) \sup_{|\zeta-z|=\varepsilon} \{ \, ||w(\zeta) - w(z)|| \, (||\phi_\zeta(\zeta)|| + ||\phi_{\overline{\zeta}}(\zeta)||) \, \} \quad .$$

Since w, ϕ_ζ , and $\phi_{\overline{\zeta}}$ are continuous this last estimate approaches zero with ε, and hence (5.5) follows from (5.6).

Corollary 10 (Cauchy Integral Formula). Let w be Q - holomorphic in Ω and continuous in $\overline{\Omega}$, where Ω is a regular subdomain of Ω_0 with boundary Γ. Then for z in Ω ,

$$w(z) = P^{-1} \int_\Gamma [\phi(\zeta) - \phi(z)]^{-1} \, d\phi(\zeta) \, w(\zeta) \quad . \tag{5.7}$$

We note that if w is Q - holomorphic in a given domain then the representation (5.7) shows that w is infinitely ϕ - differentiable in that domain, and all the ϕ - derivatives of w are also Q - holomorphic. In fact, differentiation of (5.7) with respect to ϕ gives the representation

$$\frac{d^n w}{d\phi^n}(z) = n! \, P^{-1} \int_\Gamma [\phi(\zeta) - \phi(z)]^{-n-1} \, d\phi(\zeta) \, w(\zeta) \quad . \tag{5.8}$$

Theorem 11 (Morera's Theorem). Let w be an $m \times s$ matrix valued function continuous in a domain Ω. Then w is Q - holomorphic in Ω if and only if

$$\int_{\partial R} (d\phi) \, w = 0 \tag{5.9}$$

for all rectangles R with closure contained in Ω .

Proof. If w is Q - holomorphic in Ω, then (5.9) follows from Corollary 7. Conversely, suppose w is continuous in Ω and that (5.9) holds for rectangles. If D is any disk in Ω with center z_0 , the line integral

$$v(z) \equiv \int_{z_0}^{z} d\phi(\zeta) \, w(\zeta) = \int_{z_0}^{z} [\phi_\zeta(\zeta) \, w(\zeta) \, d\zeta + \phi_{\overline{\zeta}}(\zeta) \, w(\zeta) \, d\overline{\zeta} \,]$$

taken along rectangular paths defines a function in D satisfying

$$v_z = \phi_z w \quad , \quad v_{\overline{z}} = \phi_{\overline{z}} w \quad .$$

Hence,

$$v_{\overline{z}} = Q \, \phi_z \, w = Q \, v_z \quad , \quad \frac{dv}{d\phi} = \phi_z^{-1} \, v_z = w \quad ,$$

and w is the ϕ - derivative of a Q - holomorphic function in D.

The next theorem is proved in a standard way be expanding the Cauchy kernel $[\phi(\zeta) - \phi(z)]^{-1}$ in (5.7) into a geometric series. Similar results are proved in [4], [10], and [11, 12]. Therefore the proof is omitted.

Theorem 12. Let w be Q - holomorphic in a domain Ω except possibly at an isolated point z_0. If w is also Q - holomorphic at z_0, then w may be expanded into a Taylor series,

$$(5.10) \qquad w(z) = \sum_{n=0}^{\infty} \frac{1}{n!} [\phi(z) - \phi(z_0)]^n \frac{d^n w}{d\phi^n}(z_0) ,$$

which converges uniformly in a sufficiently small neighborhood of z_0. More generally, if w has an isolated singularity at z_0, then w can be expanded into a Laurent series,

$$(5.11) \qquad w(z) = \sum_{n=-\infty}^{\infty} [\phi(z) - \phi(z_0)]^n C_n ,$$

where C_n is given by the integral

$$(5.12) \qquad C_n = P^{-1} \int_{|\zeta - z| = \rho} [\phi(\zeta) - \phi(z_0)]^{-n-1} d\phi(\zeta) w(\zeta) ,$$

and ρ is chosen so that the disk $|\zeta - z| \leq \rho$ is contained in Ω. The Laurent series converges in a sufficiently small deleted neighborhood of z_0, and convergence is uniform on compact subsets of this deleted neighborhood.

6. THE CASE OF Q CONSTANT

When Q is a constant matrix the function theory for the generalized Beltrami system

$$(6.1) \qquad w_{\bar{z}} = Q w_z$$

is particularly simple. A generating solution for (6.1) in the entire complex plane is

$$\phi(z) \equiv z I + \bar{z} Q .$$

Since Q has no eigenvalues of magnitude one, the matrix

$$\phi(z) - \phi(\zeta) = \phi(z - \zeta) = (z - \zeta)I + (\bar{z} - \bar{\zeta}) Q = (\bar{z} - \bar{\zeta})[\frac{z-\zeta}{\bar{z}-\bar{\zeta}} I + Q]$$

is invertible if $z \neq \zeta$. For a given Q we define the following positive constants γ_i, i = 1, ..., 4 :

$$\gamma_1 = \sup_{|z|=1} ||zI + \bar{z} Q|| , \qquad \gamma_2 = \sup_{|z|=1} ||(zI + \bar{z} Q)^{-1}||$$

$$\gamma_3 = \inf_{|z|=1} ||zI + \bar{z} Q|| , \qquad \gamma_4 = \inf_{|z|=1} ||(zI + \bar{z} Q)^{-1}|| .$$

For any $z \in \mathbb{C}$, $z \neq 0$, we have

$$||\phi(z)|| = ||zI + \bar{z}\,Q|| = |z|\ ||\phi(z/|z|)|| \quad .$$

Thus we have the inequality

$$\gamma_3|z| \leq ||\phi(z)|| \leq \gamma_1|z| \quad .$$

In a similar manner we obtain

$$\gamma_4|z|^{-1} \leq ||\phi(z)^{-1}\ || \leq \gamma_2|z|^{-1} \quad , \quad z \neq 0 \quad .$$

Hence we see that properties (G1) - (G6) of section 4 are easily verified for this choice of ϕ. We also have $\phi_z = I$. Thus if w is Q - holomorphic, the ϕ - derivative of w (see (4.3)) becomes

$$\frac{dw}{d\phi} = w_z \quad ,$$

and the Taylor series (5.10) becomes

$$w(z) = \sum_{n=0}^{\infty} \frac{1}{n!}\ [(z-z_0)I + (\bar{z}-\bar{z}_0)Q]^n\ \frac{\partial^n w}{\partial z^n}(z_0) \quad .$$

7. EXISTENCE OF A GENERATING SOLUTION

We now prove the existence of a generating solution for the generalized Beltrami system,

(7.1) $$w_{\bar{z}} = Q\ w_z \quad .$$

For complex valued functions g defined on the whole plane $\mathbb{C}$ we employ the usual norms,

$$||g||_\infty = \operatorname*{ess\ sup}_{z \in \mathbb{C}} |g(z)| \quad , \quad ||g||_p = \{ \int_{\mathbb{C}} |g|^p\ dx\ dy\ \}^{1/p}\ , \ 0 < p < \infty.$$

If $v = (v_{ij})$ is a complex matrix valued function on $\mathbb{C}$, we define

$$||v||_p = \sum_{i,j} ||v_{ij}||_p \quad , \quad 0 < p \leq \infty \quad .$$

The following existence theorem will be proved :

<u>Theorem 13</u>. Let Q be an mxm complex matrix which is bounded, Holder - continuous, and self-commuting in $\mathbb{C}$, and suppose that there exists an $\varepsilon > 0$ such that at each point z in $\mathbb{C}$ the eigenvalues $\lambda_1(z), \ldots, \lambda_m(z)$ of Q(z) all satisfy the condition

(7.2) $$|\ 1 - |\lambda_i(z)|\ | \geq \varepsilon \quad , \quad i = 1, \ldots, m \quad .$$

Suppose also that there exists a constant matrix Q_0 such that

$$(7.3)\qquad \begin{array}{c} Q(z) \longrightarrow Q_0 \quad \text{as } z \to \infty \\ Q - Q_0 \in L_{p'}(\mathbb{C}) \quad \text{for some } p' , \ 1 \le p' < 2 \quad . \end{array}$$

Then there exists a generating solution ϕ for (7.1) in the entire plane $\mathbb{C}$ having properties (G1) - (G4) of section 4. Moreover, ϕ has the additional properties

$$(7.4)\qquad \phi(z) - zI - \bar{z}\, Q_0 \quad \text{is bounded in } \mathbb{C}$$

$$(7.5)\qquad \phi_{\bar{z}} - Q_0 \ , \ \phi_z - I \ \text{ are in } L_p(\mathbb{C}) \text{ for all } p \text{ sufficiently close to } 2 \ ,$$

and there exist positive constants M, α, with $0 < \alpha < 1$, such that

$$(7.6)\qquad ||\phi(z_1) - \phi(z_2)|| \le M|z_1 - z_2|^{\alpha} \quad , \quad \text{for all } z_1 , z_2 \text{ in } \mathbb{C} \quad .$$

Proof. The theorem will be proved in stages. We first assume that $Q = (q_{ij})$ is an mxm complex matrix with the special form

$$(7.7)\qquad Q = \begin{bmatrix} \lambda & & & 0 \\ & \lambda & & \\ & & \ddots & \\ * & & & \lambda \end{bmatrix} \quad .$$

Thus Q is lower triangular and may be written as

$$Q(z) = \lambda(z)\, I + N(z) \quad ,$$

where N(z) is nilpotent with all nonzero terms lying below the main diagonal. We shall assume for now that the entries of Q satisfy the inequalities

$$(7.8)\qquad \sum_{i=1}^{m} ||q_{ik}||_{\infty} \le q_0 < 1 \quad , \quad k = 1, \ldots, m \quad ,$$

for some nonnegative constant q_0 . Condition (7.3) implies that there exists a constant matrix Q_0 of the form

$$Q_o = \begin{bmatrix} \lambda_o & & & 0 \\ & \lambda_o & & \\ & & \ddots & \\ * & & & \lambda_o \end{bmatrix} \quad ,$$

where $|\lambda_0| \le q_0 < 1$, which commutes with Q(z) for all z in $\mathbb{C}$, and for which (7.3) holds.

The method we use to obtain a generating solution is a modification of that of Bojarskii [3], who assumed the further condition that Q vanish identically outside some compact subset of the plane. This method makes use of the well known integral operators T and Π , defined formally by

$$(Tv)(z) = -\frac{1}{\pi}\iint_{\mathbb{C}} \frac{v(\zeta)}{\zeta - z}\, d\xi\, d\eta$$

$$(\Pi v)(z) = -\frac{1}{\pi}\iint_{\mathbb{C}} \frac{v(\zeta)}{(\zeta - z)^2}\, d\xi\, d\eta \quad .$$

The properties of these properties are discussed extensively in [16], chapter I. The properties that we will use are listed in the following lemma. For proofs the reader may consult [16].

Lemma 14. (i) If $v \in L_p(\mathbb{C}) \cap L_{p'}(\mathbb{C})$, where $1 \leq p' < 2$, $2 < p < \infty$, then Tv is bounded and uniformly Holder - continuous in $\mathbb{C}$ with exponent $\alpha = (p-2)/p$, and satisfies inequalities of the form

$$||Tv||_\infty \leq M(p,p')[||v||_p + ||v||_{p'}]$$

$$|(Tv)(z_1) - (Tv)(z_2)| \leq M(p,p')[||v||_p + ||v||_{p'}]\ |z_1 - z_2|^\alpha \ , \quad z_1,z_2 \in \mathbb{C}.$$

Moreover, in the Sobolev sense in $\mathbb{C}$ we have the differentiation formulas

$$(Tv)_{\bar{z}} = v \quad , \quad (Tv)_z = \Pi v \quad .$$

(ii) The operator Π is a bounded operator on $L_p(\mathbb{C})$ for $1 < p < \infty$, satisfying an inequality of the form

$$||\Pi v||_p \leq \Lambda_p\ ||v||_p \quad , \tag{7.9}$$

where $\Lambda_p > 0$. Moreover we have $\Lambda_2 = 1$, and given $\varepsilon > 0$, there exists $\delta > 0$ such that

$$|p - 2| < \delta \ \Rightarrow\ \Lambda_p \leq 1 + \varepsilon \quad . \tag{7.10}$$

The above properties are shown in [16] to hold for the case when v is a scalar valued function, but it is simple to see that they remain true for matrix valued functions as well.

The generating solution ϕ of (7.1) is an mxm complex matrix satisfying in $\mathbb{C}$ the equation

$$\phi_{\bar{z}} = Q\ \phi_z \quad . \tag{7.11}$$

We will find a solution ϕ having Sobolev derivatives in $L_p(\mathbb{C})$ for p close to 2. We look for a solution of (7.11) of the form

$$\phi(z) = zI + \bar{z}\ Q_0 + (Tv)(z) \quad , \tag{7.12}$$

where v is an mxm matrix in $L_p(\mathbb{C})$ for all p in some neighborhood of $p = 2$.

We define a function Q_1 by

$$Q_1(z) \equiv Q(z) - Q_0 \quad , \quad z \in \mathbb{C} \quad .$$

From (7.3) and the fact that Q is bounded it follows that $Q_1 \in L_p(\mathbb{C})$ for all $p \geq p'$. Differentiation of (7.12) yields, in the Sobolev sense,

$$(7.13) \qquad \phi_{\overline{z}} = Q_0 + v \quad , \quad \phi_z = I + \Pi v \quad .$$

Substitution into (7.11) yields the following singular integral equation for v,

$$(7.14) \qquad v = Q_1 + Q\Pi v \quad .$$

From (7.8) and (7.9) we obtain

$$||Q\Pi v||_p \leq q_0||\Pi v||_p \leq q_0\Lambda_p||v||_p \quad , \quad 0 < p < \infty \quad .$$

Since $q_0 < 1$, and (7.10) holds, the operator $Q\Pi$ is a contraction in any space $L_p(\mathbb{C})$ provided that p is sufficiently close to 2 so that the inequality

$$q_0 \Lambda_p < 1$$

holds. Thus equation (7.14) is uniquely solvable in $L_p(\mathbb{C})$ for all p in some interval of the type $[2-\varepsilon,2+\varepsilon]$, $\varepsilon > 0$. The generating solution ϕ is then given by the representation (7.12) in terms of the solution v of (7.14). It remains to verify the required properties (7.4) - (7.6) and (G1) - (G4) of the generating solution.

Properties (7.4) and (7.6) are consequences of the properties of the T operator listed in Lemma 14, and (7.5) is a consequence of (7.13) and the fact that v, Qv are in $L_p(\mathbb{C})$ for p close to 2. Since Q is Holder - continuous, any solution ϕ of (7.11) has Holder - continuous first partial derivatives. Hence (G1) holds, and moreover, from (7.13) we conclude that the functions v, Πv are also Holder - continuous in $\mathbb{C}$.

To verify property (G2) we note that since $Q\Pi$ is a contraction the solution v of (7.14) has the expansion

$$(7.15) \qquad v = \sum_{k=0}^{\infty} (Q\Pi)^k \, Q_1 \quad ,$$

which converges in the $L_p(\mathbb{C})$ sense if p is close enough to 2. From this representation we conclude that

$$v(z_1)\, Q(z_2) = Q(z_2)\, v(z_1) \quad , \quad v(z_1)\, v(z_2) = v(z_2)\, v(z_1) \quad , \quad z_1, z_2 \in \mathbb{C} \; .$$

(We know first of all that these relations hold only in the $L_p(\mathbb{C})$ sense, and hence almost everywhere, but since the functions are continuous they must hold point-wise as well.) From the representation (7.12) for ϕ and the properties of the T operator it follows then that ϕ also is self-commuting and commutes with Q in $\mathbb{C}$.

It remains to verify properties (G3) and (G4) of the generating solution. From the expansion (7.15) and the representation (7.12) we see that v and ϕ are lower triangular matrices of the form

$$(7.16)\qquad v = \begin{bmatrix} v_0 & & & 0 \\ & v_0 & & \\ & & \ddots & \\ * & & & v_0 \end{bmatrix}, \qquad \phi = \begin{bmatrix} \phi_o & & & 0 \\ & \phi_o & & \\ & & \ddots & \\ * & & & \phi_o \end{bmatrix}.$$

The diagonal term ϕ_0 of ϕ satisfies an ordinary Beltrami equation for scalar valued functions,

$$(7.17)\qquad \phi_{0,\bar{z}} = \lambda\, \phi_{0,z} ,$$

and has the representation

$$\phi_0(z) = z + \bar{z}\,\lambda_0 + (Tv_0)(z) .$$

Since Tv_0 is bounded, and $|\bar{z}\,\lambda_0| \le q_0|z|$, the change in the argument of $\phi_0(z)$ as z traces in a positive direction a sufficiently large circle about the origin is given by

$$\Delta\,[\arg \phi_0(z)] = \Delta(\arg z) = 2\pi .$$

Since the argument principle holds for solutions of Beltrami equations (7.17) whenever λ is Holder - continuous (see [16], chapter II), we conclude that ϕ_0 is a one - to - one mapping on $\mathbb{C}$. Hence,

$$\phi_0(z) - \phi_0(\zeta) \ne 0 , \quad z, \zeta \in \mathbb{C} , z \ne \zeta .$$

From the special triangular form (7.16) of ϕ it follows that $\phi(z) - \phi(\zeta)$ is invertible if $z \ne \zeta$, and therefore (G3) holds. Furthermore, it follows from properties of one - to - one solutions of single Beltrami equations that $\phi_{0,z} \ne 0$ in $\mathbb{C}$ (see [16], chapter II). Hence $\phi_z(z)$ is invertible for all points z in $\mathbb{C}$, and (G4) holds.

We next weaken the conditions on Q by replacing (7.8) with the simpler condition

$$(7.18)\qquad |\lambda(z)| \le q_0 < 1 , \quad z \in \mathbb{C} .$$

We still assume that Q has the form (7.7) and that the entries of Q are bounded. Let δ be a positive constant, and let D be the mxm constant and diagonal matrix,

$$D = \begin{bmatrix} 1 & & & & 0 \\ & \delta & & & \\ & & \delta^2 & & \\ & & & \ddots & \\ 0 & & & & \delta^{m-1} \end{bmatrix}.$$

Let $\tilde{Q} = (\tilde{q}_{ij})$ be the mxm matrix

$$\tilde{Q} \equiv DQD^{-1} \quad .$$

Then $\tilde{Q}$ also has the form (7.7) with the same diagonal term λ. However, the nondiagonal terms of $\tilde{Q}$ are $\mathcal{O}(\delta)$ as $\delta \longrightarrow 0$. Thus for δ sufficiently small the entries of $\tilde{Q}$ satisfy an inequality of the form

$$\sum_{i=1}^{m} ||\tilde{q}_{ik}||_{\infty} \le q_0 < 1 \quad , \quad k = 1, \ldots, m \; .$$

Hence there exists a generating function ψ corresponding to the matrix $\tilde{Q}$ which satisfies the equation

$$\psi_{\overline{z}} = \tilde{Q}\, \psi_z = DQD^{-1}\psi_z \quad ,$$

along with corresponding conditions (G1) - (G4), (7.4) - (7.6) (with Q replaced by $\tilde{Q}$). It is easily verified then that the function

$$\phi \equiv D^{-1}\, \psi\, D$$

is a generating solution corresponding to the matrix Q which satisfies all the required conditions.

As the next step in the proof of Theorem 13 we consider the case where Q again has the form (7.7), but instead of (7.18) we assume that the diagonal term λ satisfies

$$(7.19) \qquad |\lambda(z)| \ge r_0 > 1 \quad , \quad z \in \mathbb{C} \quad ,$$

for some constant r_0. Hence the limit matrix Q_0 of Q(z) at infinity, appearing in (7.3), is invertible. We define a function ψ in terms of the required generating solution ϕ for Q by

$$\psi(z) \equiv {Q_0}^{-1}\, \phi(\overline{z}) \quad .$$

Then from the equation

$$\phi_{\overline{z}} = Q\, \phi_z$$

for ϕ we obtain the equation

$$\psi_{\overline{z}}(z) = \tilde{Q}(z)\, \psi_z(z)$$

involving ψ, where now $\tilde{Q}$ is defined by

$$\tilde{Q}(z) \equiv Q(\overline{z})^{-1} \quad .$$

Writing

$$Q(z) = \lambda(z)\, I + N(z) \quad ,$$

where N(z) is nilpotent, we obtain for $\tilde{Q}(z)$ the expansion

(7.20)
$$\tilde{Q}(z) = \sum_{k=0}^{m-1} (-1)^k \lambda(\overline{z})^{-k-1} N(\overline{z})^k \quad .$$

Thus $\tilde{Q}$ has the form (7.7), but with the diagonal term $\lambda(z)$ replaced by the quantity $\lambda(\overline{z})^{-1}$, which satisfies the inequality

$$|\lambda(\overline{z})^{-1}| \leq 1/r_0 < 1 \quad .$$

Moreover, since Q is bounded, Holder - continuous, and self - commuting, the representation (7.20) implies that $\tilde{Q}$ also has these properties. We also have the identity

$$\tilde{Q}(z) - Q_0^{-1} = \tilde{Q}(z)\, Q_0^{-1} [Q_0 - Q(\overline{z})] \quad .$$

From this identity and from the behaviour of Q at infinity, given by (7.3), we deduce that the behaviour of $\tilde{Q}$ at infinity is described by

$$\tilde{Q}(z) \longrightarrow Q_0^{-1} \quad \text{as } z \rightarrow \infty \quad ,$$

$$\tilde{Q} - Q_0^{-1} \in L_{p'}(\mathbb{C}) \quad \text{for some } p', \ 1 \leq p' < 2 \quad .$$

Hence $\tilde{Q}$ satisfies conditions sufficient to imply the existence of a generating solution ψ. Let ψ be the generating solution, corresponding to $\tilde{Q}$, which satisfies (G1) - (G4) and (7.4) - (7.6) with Q replaced by $\tilde{Q}$ and Q_0 replaced by Q_0^{-1}. It is not hard to verify then that the function ϕ given by

$$\phi(z) = Q_0\, \psi(\overline{z})$$

is a generating solution for Q.

As the final step in the proof of Theorem 13, we consider the general case where Q satisfies only the hypotheses of the theorem. We make a transformation of the type

$$\hat{Q}(z) \equiv S\, Q(z)\, S^{-1} \quad ,$$

as described in section 2, where S is a constant matrix and $\hat{Q}$ has the canonical form of (2.6) and (2.7). Then (7.1) becomes

(7.21)
$$v_{\overline{z}} = \hat{Q}\, v_z$$

where $v \equiv Sw$. Then the system separates into subsystems, each of which is of a type for which it has already been proved that a generating solution exists. By combining these generating solutions for the individual subsystems into one matrix we obtain a generating solution ψ for (7.21). Then $\phi \equiv S^{-1} \psi S$ is a generating solution for (7.1).

ACKNOWLEDGEMENTS

The author was supported in part by National Science Foundation Grant MCS 78 - 01993 and by a fellowship from the Humboldt Foundation of West Germany, while visiting at the Technische Hochschule in Darmstadt, Federal Republic of Germany.

REFERENCES

1. H. Begehr and R. P. Gilbert, On Riemann boundary value problems for certain linear elliptic systems in the plane, J. Differential Equations 32 (1979), 1 - 14.
2. L. Bers, Theory of Pseudo-analytic Functions, N. Y. U. Lecture Notes, 1953.
3. B. Bojarskii, Die Theorie des verallgemeinerten analytischen Vektors (Aufsatz in russicher Sprache, Übersetz aus dem Russischem Bibl. der Tech. Univ. Hannover). Anns. Pol. Math. 17 (1966), 281 - 320.
4. A. Douglis, A function theoretic approach to elliptic systems of equations in two variables, Comm. Pure Appl. Math VI (1953), 259 - 289.
5. A. Douglis and L. Nirenberg, Interior estimates for elliptic systems of partial differential equations, Comm. Pure Appl. Math VIII (1955), 503 - 538.
6. R. P. Gilbert, Constructive Methods for Elliptic Equations, Springer - Verlag, Heidelberg, 1974.
7. R. P. Gilbert, Pseudo hyperanalytic function theory, Proceedings of the Conference "Funktionentheoretische Eigenschaften von Lösungen partieller Differentialgleichungen", ed. by St. Ruscheweyh, 1974.
8. R. P. Gilbert and G. N. Hile, Generalized hypercomplex function theory, Trans. Amer. Math. Soc. 195 (1974), 1 - 29.
9. R. P. Gilbert and G. N. Hile, Hypercomplex function theory in the sense of L. Bers, Math. Nachr. 72 (1976), 187 - 200.
10. B. Goldschmidt, Funktionentheoretische Eigenschaften verallgemeinerter analytischer Vektoren, Math. Nachr. 90 (1979), 57 - 90.
11. G. N. Hile, Elliptic systems in the plane with first order terms and constant coefficients, Comm. Partial Differential Equations 3 (1978), 949 - 977.
12. G. N. Hile, Function theory for a class of elliptic systems in the plane, J. Differential Equations 32 (1979), 369 - 387.
13. G. N. Hile and W. E. Pfaffenberger, The relative spectrum of elements in a real Banach algebra, Math. Annalen 250 (1980), 113 - 133.

14. N. Jacobson, Lectures in Abstract Algebra, II, Linear Algebra, D. Van Nostrand, Princeton, 1953.

15. E. Kühn, Über die Funktionentheorie einer Klasse elliptischer Differentialgleichungssysteme in der Ebene, Dissertation, Dortmund, 1974.

16. I. N. Vekua, Generalized Analytic Functions, Addison - Wesley, Reading, 1962.

Contemporary Mathematics
Volume 11, 1982

ON A VARIATIONAL INEQUALITY FOR THE HODOGRAPH METHOD

Robert A. Hummel

1. Planar fluid flow

The hodograph method has been studied extensively for incompressible and compressible inviscid irrotational planar fluid flow. The principal advantage of the hodograph method is that the system of differential equations become linear when expressed in terms of the hodograph variables. The major disadvantages are that the hodograph transformation may not be one-to-one, and that the boundaries of the hodograph domain are unknown. Recently, the use of variational inequalities has been shown to overcome these difficulties in certain cases. In this paper, we outline the current status of these results.

The equations of motion for planar, inviscid, irrotational fluid flow are

$$\operatorname{div} \rho(|\bar{q}|)\, \bar{q} = 0 \qquad (1)$$

$$\operatorname{curl} \bar{q} = 0 \qquad (2)$$

where $\rho = \rho(|q|)$ is a given positive decreasing function,

and $\bar{q} = \bar{q}(x,y) = (q_1(z), q_2(z))$, $z = x+iy$.

Equation (1) is the "equation of continuity", and expresses the physical property of conservation of matter. The second equation is unphysical, since it results from a theorem whose hypotheses include a perfect fluid model of the material. As a special case, incompressible fluid flow results when $\rho(q)$ is simply a positive constant, instead of a given decreasing function. The usual model for $\rho(q)$ for a compressible fluid is obtained from the Bernoilli

0271-4132/82/0000-0145/$03.75

relation (whose derivation uses (2)), $|q|^2/2 + P/\rho = \text{const.}$, and an adiabatic pressure-density law, such as $p = c\rho^{\gamma}$.

It is not hard to show that the system (1)-(2) is elliptic whenever $M = M(q) = q/c < 1$, where

$$c^2 = - \frac{\rho \cdot q}{\rho'(q)} \tag{3}$$

For incompressible flow, the system is always elliptic. For a general density-speed relation $\rho(q)$, $M < 1$ for sufficiently small q. For ideal fluids, one can show that M is increasing in q, and that there is exactly one critical speed q^* satisfying $M(q^*) = 1$. Accordingly, the system (1)-(2) is elliptic providing the solution $\overline{q}$ is subsonic, i.e., $|\overline{q}| < q^*$ everywhere.

The ellipticity of the system (1)-(2) is equivalent to the statement that

$$V(z) = q_1(z) - iq_2(z)$$

is quasi-conformal. Indeed, for the incompressible case, we may take $\rho \equiv 1$ without loss of generality, in which case (1) and (2) are simply the Cauchy-Riemann equations for $V(z)$. The hodograph transformation replaces the physical coordinate z of the flow region with the coordinate $\zeta = V(z)$ formed by the components of velocity. If the map $z \to V(z)$ is locally one-to-one at a point, then the flow solution near that point can be represented by the inverse map, which is a function of the hodograph variable $\zeta = q_1 - iq_2$. For subsonic flow, since $V(z)$ is quasi-conformal, the singularities where the hodograph transformation is not one-to-one are discrete and well behaved.

The hodograph method using variational inequalities, as investigated by Brezis and Stampacchia [1] can treat a variety of planar fluid flow problems. Most of the work has concentrated on exterior domains - flow past an obstacle with prescribed velocity at infinity. However, flow in a channel with an obstacle, flow through a Lavalle nozzle, and flow past an obstacle with cavitation can all be treated by similar methods [2,3,4] . To date, these methods always require a convexity condition limiting the kinds of geometry of the profile and walls that can be treated. For example, for flow past an obstacle, we require the obstacle to be strictly convex. We will also require that the solution satisfy a zero-circulation condition. Because of the perfect fluid assumption, flow past an obstacle exhibits no drag, and there is a one parameter family of solutions, parameterized by

$$\Gamma = \int_{\partial P} q_1 dx + q_2 dy \ ,$$

where P is the obstacle profile. Variational methods in conjunction with the hodograph transform have dealt only with the case $\Gamma = 0$. Finally, we note that hodograph methods almost necessarily are restricted to planar fluid flow situations, whether or not variational inequalities are employed.

2. Flow past an obstacle

We restrict our discussion to the problem of finding the flow past a convex profile with prescribed velocity at infinity. We are given,

(i) $P \subset \mathbb{C}$, bounded, strictly convex, $C^{2,\infty}$,

(ii) q_∞ , $0 < q_\infty \ll q^*$, and

(iii) $\rho = \rho(q)$, density-speed relation of an ideal fluid, or $\rho \equiv 1$.

Defining the flow region $G = \mathbb{C} \setminus P$, the flow problem can be stated as

Find $\bar{q} = (q_1, q_2) \in [C^2(G) \cap C^0(G)]^2$, such that

$$\operatorname{div} \rho(|q|)\bar{q} = 0 \quad \text{in } G \tag{2.1}$$

$$\operatorname{curl} \bar{q} = 0 \quad \text{in } G \tag{2.2}$$

$$\bar{q} \cdot \bar{n} = 0 \quad \text{on } G , \tag{2.3}$$

$$\bar{n} = \text{normal to } \partial P,$$

$$\bar{q}(z) \to (q_\infty, 0) \text{ as } |z| \to \infty \tag{2.4}$$

and

$$\Gamma = \int_{\partial} q_1 dx + q_2 dy = 0 \tag{2.5}$$

Existence and uniqueness of a subsonic solution to (2.1-2.5) are known [5] . Variational methods using the hodoraph transform may lead to an independent existence theorem, but the principal motivation is the simple numerical implementation of a variational inequality.

Associated with the solution $\bar{q}$ are the potential and stream functions

$$\phi_x = q_1 \ , \ \phi_y = q_2$$

$$\Psi_x = -\rho q_2 \ , \ \Psi_y = \rho q_1 \ .$$

The stream function Ψ is constant on streamlines, so that we may assume that $\Psi \equiv 0$ on ∂P. Eliminating ϕ, the stream function satisfies the quasilinear equation

$$(1 - \frac{q_1^2}{c^2})\Psi_{xx} - 2\,\frac{q_1 q_2}{c^2}\,\Psi_{xy} + (1 - \frac{q_2^2}{c^2})\Psi_{yy} = 0$$

If Ψ is known, q can be recovered using $\text{grad}\Psi = (-\rho q_2, \rho q_1)$, and the fact that $\rho(q)\cdot q$ is an increasing, and therefore invertible function of q, for $q < q^*$.

3. The Hodograph Domain

Since the function V(z) is not globally one-to-one, some points in the hodograph domain may be multiply covered. A key idea in the extension of hodograph methods to non-symmetric flow is to view the hodograph domain as a Riemann surface, where points which are covered more than once are separated into multiple sheets. When the profile is convex, the Riemann surface can be described a-priori.

The description of the Riemann surface depends on the functions chosen to coordinatize the sheets. Of course, the coordinates (q_1, q_2) are the most obvious choice, although $(\rho(q) \cdot q_1, \rho(q) \cdot q_2)$ can also serve. In hodograph methods, it is common to use a polar-coordinate type representation (θ, σ), where

$$\theta = \arg(q_1 + iq_2) \tag{3.1}$$

$$\sigma = \int_{|q|}^{q^*} \frac{\rho(s)}{s}\, ds\ . \tag{3.2a}$$

For incompressible flow, the definition

$$\sigma = -\log |q| \tag{3.2b}$$

must be used in place of (3.2a). Note that $\sigma = +\infty$ corresponds to a stagnation point $q_1 - iq_2 = 0$, whereas for compressible flow, $\sigma = 0$ is the sonic speed.

To derive a description of the Riemann surface in the (q_1, q_2) variables, one first observes that by the classical existence theory, the solution $V(z)$ is branched over $z = \infty$, where $V(\infty) = q_\infty$. Further, one can show that $V(z) = 0$ at exactly two points, both located on the profile boundary. At a point z on the profile boundary, $\overline{V(z)}$ lies in the same direction as a tangent to ∂P at z. Thus $V(\partial P)$ consists of two closed curves, joined at the origin, and both enclosing q_∞ in their bounded interior regions. Consequently, the Riemann surface consists of two sheets, with a first order branch point at $\zeta = q_\infty$; each sheet is bounded by one of the two closed curves (see Figure 1). It is convenient to view the surface cut along the line $0 \leq \zeta < q$. The two sheets are joined along

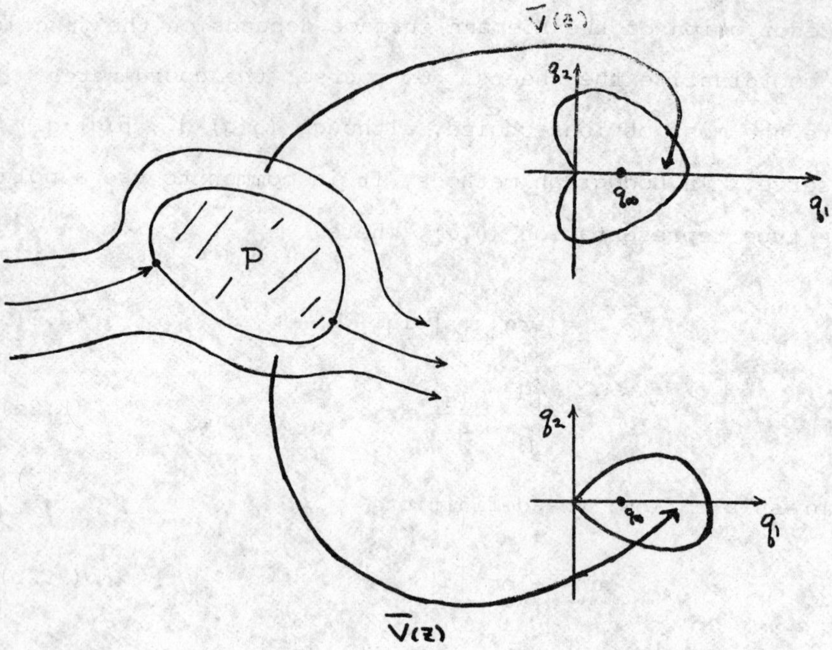

Figure 1.

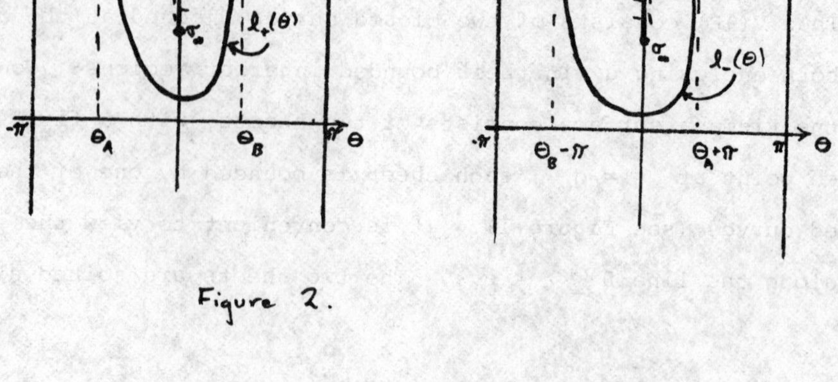

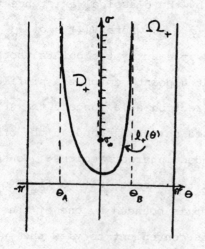

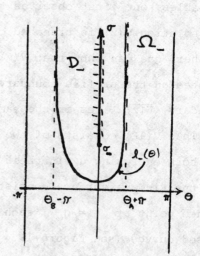

Figure 2.

the cut line in a criss-cross fashion, so that the upper shore on either sheet is identified with the lower shore on the other sheet.

In terms of the (θ,σ) variables, each sheet lies inside strip domains $\{(\theta,\sigma) = -\pi \leq \theta \leq \pi, \sigma \geq 0\}$. If we denote the two sheets by D_+ and D_-, we have

$$D_\pm = \{(\theta,\sigma) : \sigma \geq \ell_\pm(\theta)\} \setminus \{(0,\sigma) : \sigma \geq \sigma_\infty\} ,$$

where $\sigma_\infty = \sigma(q_\infty)$ is the σ-value at the branch point. The curves $\ell_+(\theta)$ and $\ell_-(\theta)$ correspond to $\sigma(q)$ along the boundary of the profile in the physical plane, and constitute an unknown free boundary of the hodograph domain. Each curve has a left and right vertical assymptote at values of θ that correspond to the (unknown) locations of the stagnation points on ∂P. The hodograph domain is the Riemann surface D consisting of D_+ and D_-, branched over $(0,\sigma_\infty)$, and identified along the cuts left shore to right and right shore to left (see Figure 2). By viewing the hodograph domain as a Riemann surface, we obtain a hodograph transform of $G \cup \{\infty\}$ onto D which is globally one-to-one.

Finally, D can be considered as a subset of a Riemann surface Ω, whose two sheets Ω_+ and Ω_- both consist of strip domains without the slits, with the same identifications on the slits as exist in D. The larger Riemann surface Ω will be the domain of the competing functions in the variational approach considered in the next section.

4. Flow equations in the hodograph variables

In the hodograph variables (θ,σ), Chapyglin's equation states that the stream function satisfies

$$k(\sigma)\,\frac{\partial^2\Psi}{\partial\theta^2} + \frac{\partial^2\Psi}{\partial\sigma^2} = 0$$

where
$$k(\sigma) = \frac{1-M^2}{\rho^2}$$

(See [6] for a derivation.)

For the variational method indicated below, we use the Legendre transform of Ψ , defined by

$$\overline{\Psi} = \Psi - x\,\frac{\partial\Psi}{\partial x} - y\,\frac{\partial\Psi}{\partial y}\ .$$

Note that $\overline{\Psi}$, as well as Ψ, can be considered to be defined in the hodograph domain D. It can be shown that Legendre transform $\overline{\Psi}$ satisfies the differential equation

$$\frac{1}{\rho^2}\,\frac{\partial^2\overline{\Psi}}{\partial\theta^2} + \frac{\partial}{\partial\sigma}\left(\frac{1}{k(\sigma)\rho^2}\,\frac{\partial\overline{\Psi}}{\partial\sigma}\right) = 0 \tag{4.1}$$

However, $\overline{\Psi}$ is non-zero on ∂D, and so we instead consider

$$U = \overline{\Psi} - q\rho(X_{\pm}(\theta)\,\sin\theta - Y_{\pm}(\theta)\,\cos\theta)\ , \tag{4.2}$$

where $(X_{\pm}(\theta),\ Y_{\pm}(\theta))$ is the coordinate of ∂P whose clockwise (resp, counter-clockwise) oriented tangent has argument θ . The function U, defined on D, does vanish on ∂D, and in fact satisfies gradU = 0 on ∂D. We extend U to be zero in the strip domains Ω_+ and Ω_- outside D_+ and D_-. This function satisfies a variational inequality.

<u>Theorem</u>: Let $\mathbb{K}$ = {V defined on the Riemann surface Ω , denoted by $V_+(\theta,\sigma)$ on Ω_+, and $V_-(\theta,\sigma)$ on Ω_-, satisfying

i. $V_{\pm} \in H^1(\Omega_{\pm})$

ii. $V_{\pm}(-\pi,\sigma) = V_{\pm}(\pi,\sigma) = 0 \qquad \forall\sigma$

iii. $\gamma V_+(0^+,\sigma) - \gamma V_-(0^-,\sigma) = q(\sigma)\cdot\rho(\sigma)\cdot H$ and
$\gamma V_+(0^-,\sigma) - \gamma V_+(0^+,\sigma) = q(\sigma)\cdot\rho(\sigma)\cdot H$, where γ is the trace operator, for $\sigma > \sigma_\infty$. (These are the "jump conditions" across the cuts.)

iv. $V_+ \geq 0$ on Ω_+, $V_- \leq 0$ on Ω_-}

Then U defined by (4.2) satisfies

$$U \in \mathbb{K}\ ,\ a(U,V-U) \geq \langle T,V-U\rangle \quad \text{for all} \quad V \in \mathbb{K}\ , \tag{4.3}$$

where $$a(u,\zeta) = \iint_{\Omega_+ \cup \Omega_-} \frac{1}{\rho^2} u_\theta \zeta_\theta + \frac{1}{k\rho^2} u_\sigma \zeta_\sigma \, d\theta d\sigma\ ,$$

and $$\langle T,\zeta\rangle = \iint_{\Omega_+ \cup \Omega} q/\rho\ R(\theta)\cdot\zeta(\theta,\sigma)d\theta d\sigma$$
$$+ \int_{\sigma_\infty}^{\infty} q/\rho \cdot W \cdot (\gamma\zeta_+(0^+,\sigma) - \gamma\zeta_+(0^-,\sigma))\ d\sigma\ .$$

Here $H = Y_+(0) - Y_-(0)$, $W = X_+(0) - X_-(0)$,

and $R_\pm(\theta) = X_\pm'(\theta)\cos\theta + Y_\pm'(\theta)\sin\theta$.

Remarks: The proof of the variational inequality (4.3) is quite simple, and follows from (4.1), once it is shown that $U \in \mathbb{K}$. Of the conditions (i)-(iv) to be satisfied by a function in $\mathbb{K}$, only condition (iv) is difficult to verify to show that U belongs to $\mathbb{K}$. Verifying (iv) requires the introduction of a class of quasi-variational inequalities, and requires a monotonicity result. The details for the incompressible case are given in [7]. The compressible case is no more difficult.

The theorem is valid for both compressible and incompressible flow. In the incompressible case, one has $\rho \equiv 1$, $k(\sigma) \equiv 1$, and $q = e^{-\sigma}$. Note that in the compressible case, $\sigma \geq 0$ by definition, whereas for incompressible flow, an a-priori lower bound for σ also exists [1] , but may be negative. When the special condition of symmetry of the profile (with respect to the horizontal axis) is imposed, the variational inequality can be shown to reduce to the ones considered in earlier work.

5. Practical Consequences

Although the variational inequality (4.3) involves functions defined on a Riemann surface, and a fairly complicated distribution T, it nontheless leads to an extremely simple algorithm for finding flow solutions to the problem (2.1)-(2.5). The variational inequality may be solved numerically to yield U, or equivalently $\bar{\Psi}$, which is composed of two functions, Ψ_+ defined in D_+, and Ψ_- defined on D_-. Note that the subset $D \subseteq \Omega$ is determined from the solution U as the set where $U \neq 0$. Viewing $\bar{\Psi}_\pm$ as functions defined in $(w_1, w_2) = (\rho q_1, \rho q_2)$ variables, simple properties of the Legendre transform show that

$$\left(\frac{\partial}{\partial w_2} - i \frac{\partial}{\partial w_1}\right)\bar{\Psi}_\pm = x + iy = z ,$$

where $V(z) = q_1 - iq_2$. That is, from $\bar{\Psi}$ one can determine the physical point assigned to a given velocity.

To find $\bar{\Psi}$, the inequality (4.3) can be formulated in complementarity form. Namely, $\bar{\Psi}$ is a smooth function on Ω, satisfies the differential equation (4.1), and satisfies the constraints

$$\bar{\Psi}_+ \geq q\rho(X_+(\theta)\cdot\sin\theta - Y_+(\theta)\cdot\cos\theta) \quad \text{on } \Omega_+$$

$$\bar{\Psi}_- \leq q\rho(X_-(\theta)\cdot\sin\theta - Y_-(\theta)\cdot\cos\theta) \quad \text{on } \Omega_- .$$

One can regard the problem as a coupled system involving $\bar{\Psi}_+$ and $\bar{\Psi}_-$, with the coupling expressed as continuity across the identifed cuts.

The use of variational inequalities for the hodograph method seems to be a natural way of locating the unknown boundary of D. The Riemann surface appears in a natural fashion also, as the image of a hodograph transform which has been made one-to-one. The convexity requirement for P has been used in the representation of the boundary, and more importantly, to specify the general structure of the Riemann surface, so that the competing functions may be defined on a fixed Riemann surface Ω.

References

1. Brezis, H., and G. Stampacchia, "The hodograph method in fluid-dynamics in the light of variational inequalities," Arch. Rat. Mech. & Analysis 61, (1976), p. 1-18.
2. Tomarelli, F., "Un probléme de fluidynamique avec les inequations variationelles," C.R. Acad. Sci. Paris 286 (1978), p999-1002.
3. Shimborsky, E., "Variational methods applied to the study of symmetric flows in laval nozzles," (Preprint).
4. Brezis, H. and G. Duvaut, "Ecoulement avec sillage autour d'un profile symétrique sans incidence," C.R. Acad. Sci. Paris 276 (1973), p875-878.
5. Bers, L., "Existence and uniqueness of a subsonic flow past a given profile," C.P.A.M. 7 , (1954), p441-504.
6. Bers, L., Mathematical Aspects of Subsonic and Transonic Gas Dynamics, Chapman and Hall, 1958.
7. Hummel, R., "The hodograph method for convex profiles," submitted to Annali de Pisa, (preprint available from the author).

Contemporary Mathematics
Volume 11, 1982

Nonlinear Boundary Value Problems of Riemann-Hilbert Type

Heinrich Begehr* and George C. Hsiao**

ABSTRACT

This paper is concerned with boundary value problems for the first-order semilinear elliptic systems of two unknowns in the plane. In particular, the nonlinear boundary conditions with negative index are considered. Existence and uniqueness of the classical solution to the problem are established for a restricted class of boundary data and the nonlinearities in the equations. The proof involves an a priori estimate and an embedding method similar to the one used for treating semilinear elliptic systems with linear boundary conditions.

1. Introduction

In recent papers [2], [3] and [5], investigations were made by the authors on the nonlinear boundary value problems for the elliptic systems of first order equations in the complex plane. In particular, existence and uniqueness results were established for the solutions to the nonlinear Cauchy-Riemann equation with nonlinear boundary conditions. The approach used in these investigations is an embedding method combined with the Newton iteration procedure [13], [15], [16]. In this paper, based on the same approach, we shall generalize the work of [2], [3] to the semilinear Riemann-Hilbert boundary value problem.

We consider here the nonlinear equation of the form***:

(E) $$w_{\bar z} = q_1(z)w_z + q_2(z)\overline{w}_z + h(z,w) \quad \text{in} \quad D,$$

*I. Mathematische Institut , Freie Universität Berlin, Hüttenweg 9, D-1000 Berlin 33, Germany.

**Department of Mathematical Sciences, University of Delaware, Newark, Delaware 19711, U.S.A. The research of this author was supported by the Freie Universität Berlin.

***Here and in the sequel, it is understood that for simplicity we omit the complex conjugate variables in the arguments of the functions, e.g. $w(z)$ and $h(z,w)$ stand for $w(z,\bar z)$ and $h(z,\bar z,w,\bar w)$ respectively.

0271-4132/82/0000-0146/$04.75

where D is a bounded, simply connected domain with a smooth boundary Γ in the complex plane $\mathbb{C}$. The nonlinear term h and the coefficients q_i are complex-valued functions satisfying certain regularity conditions to be specified later. Throughout the paper, the uniform ellipticity is assumed; that is, there exists a constant $q < 1$ such that

$$(Q) \qquad |q_1(z)|+|q_2(z)| \le q$$

holds for all $z \in \hat{D} = D \cup \Gamma$. In the special case when $q_1 = q_2 = 0$, (E) becomes the semilinear Cauchy-Riemann equation which has been treated in [2] and [3], while for $q_2 = 0$, we have the semilinear Beltrami system [7]. Following [2] and [3], we require w to satisfy the nonlinear boundary condition:

$$(B) \qquad \text{Re}\ \{e^{i\tau}w\} = \psi(z,w) \quad \text{on} \quad \Gamma,$$

where τ and ψ are given real-valued functions. We assume that the <u>index</u> n defined by

$$n := \frac{1}{2\pi}\int_\Gamma d\tau = \frac{1}{2\pi}\int_\Gamma d(\arg e^{i\tau})$$

is <u>negative</u>. Then in order to insure the uniqueness of the solution, we impose additional nonlinear side conditions

$$(C) \qquad \Sigma^{-1}\int_\Gamma \text{Im}\ \{e^{i\tau}w\}\sigma\ ds = \kappa(w) \quad \text{and}$$

$$(C') \qquad w(z_k) = a_k(w), \quad k = 1,2,\ldots,|n|,$$

where $z_k \in D$ are distinct points which are chosen arbitrarily but fixed, and $\sigma > 0$ is a given real weight function on Γ such that $\Sigma := \int_\Gamma \sigma(s)\ ds \neq 0$; κ and a_k are respectively given real- and complex-valued functional. In these formulations, the given data ψ, κ and a_k as well as the nonhomogeneous term h all depend on the unknown w in appropriate function spaces which will be made precise in later sections.

Throughout the paper we shall refer to the boundary value problem defined by (E), (B), (C) and (C') as the <u>semilinear Riemann-Hilbert boundary value problem</u>. We remark that in the case $n = 0$, condition (C') is not needed and that our analysis for $n < 0$ in this paper can be carried out in an exact way for this case. Needless to say, the semilinear Riemann-Hilbert boundary value problem may also be treated by the other methods. However, we believe that our approach is more constructive. For other methods such as methods based on singular integral equations and the Schauder-fixed point argument for treating boundary value problems of this type we refer to the work in [1], [7-12], [14] and [18].

2. An Imbedding Method

The existence proof used in [2] and [3] is based on a variant of the imbedding method of [13] and [15]. The essence of the method is to imbed the nonlinear boundary value problem under consideration into a family of problems (P_t) depending on a parameter $t \in [0,1]$ such that (P_1) is the corresponding nonlinear problem under consideration, and (P_o) is a linear problem (or a problem which is easily solvable and even possesses a known solution in suitable function space). Then the basic idea consists of continually transforming the solution $w(z,t)$ of (P_t) with $t = 0$ into the desired solution $w(z,1)$ of (P_1) in appropriate function space.

For the semilinear Riemann-Hilbert boundary value problem, (E), (B), (C) and (C'), we consider the following imbedding:

$$(P_t)\qquad \begin{aligned} &Lw = t\, h(z,w) && \text{in } D \\ &\mathrm{Re}\,\{e^{i\tau}w\} = t\,\psi(z,w) && \text{on } \Gamma \\ &\Sigma^{-1}\int_\Gamma \mathrm{Im}\,\{e^{i\tau}w\}\sigma\, ds = t\,\kappa(w) && \\ &w(z_k) = t\, a_k(w), && k = 1,2,\dots,|n|, \end{aligned}$$

where $t \in [0,1]$. Here the operator L is defined by

$$Lw := w_{\bar z} - q_1(z)w_z - q_2(z)\overline{w}_z. \tag{2.1}$$

The original problem, (E), (B), (C) and (C'), corresponds to (P_1). For $t = 0$, $w(z,0) = 0$ is the solution of (P_o). Using this solution as the initial approximation, we may solve (P_t) for $t = t_1 > 0$ by an iteration procedure which leads to a sequence of linear problems. The known solution $w(z,t_1)$ can then be used again as the initial approximation for $w(z,t_2)$ with $t_2 > t_1$. Repeating the process in this way, one expects that after finitely many steps, the desired solution $w(z,1)$ will be reached.

To describe the iteration procedure let us for the time being assume that the coefficients and the prescribed data are sufficiently smooth so that the following iteration process is meaningful. We begin by assuming that $w(z,t_{j-1})$ is the solution of (P_t) for a fixed t_{j-1}, $0 \leq t_{j-1} < 1$. Then as an approximation to the solution of (P_t) for $t = t_j$, $t_{j-1} < t_j \leq 1$, we define the sequence $\{w_n(z,t_j)\}$ by

$$w_0(z,t_j) = w(z,t_{j-1}) \qquad (z \in \hat{D}) \tag{2.2}$$

and by solving the linear boundary value problems below for $w_{m+1}(z,t_j)$, $m = 0,1,\ldots$:

$$\begin{aligned} Lw_{m+1} &= t_j\{h(z,w_m)+h_w(z,w_m)(w_{m+1}-w_m) \\ &\qquad +h_{\overline{w}}(z,w_m)(\overline{w_{m+1}}-\overline{w_m})\} \qquad \text{in} \quad D \\ \text{Re}\,\{e^{i\tau}w_{m+1}\} &= t_j\psi(z,w_m) \qquad \text{on} \quad \Gamma \\ \Sigma^{-1}\int_\Gamma \text{Im}\,\{e^{i\tau}w_{m+1}\}\sigma\, ds &= t_j\kappa(w_m) \\ w_{m+1}(z_k,t_j) &= t_j a_k(w_m), \qquad k = 1,2,\ldots,|n|. \end{aligned} \tag{2.3}$$

We note that as in [13] the linear differential equation in (2.3) is exactly the same as those obtained from (P_{t_j}) by Newton's method. Indeed, if we rewrite $Lw = t\,h(z,w)$ in the form $H(w;t) := Lw-t\,h(z,w) = 0$, then Newton's method yields the iteration:

$$H(w_m;t_j)+H'(w_m;t_j)(w_{m+1}-w_m) = 0$$

where H' denotes the Fréchet derivative of H defined by

$$H'(v;t_j)w = Lw-t_j\{h_w(z,v)w+h_{\overline{w}}(z,v)\overline{w}\}.$$

On the other hand, the iterations from the nonlinear boundary and side conditions in (2.3) are not the same as those obtained by Newton's method. Here the corresponding Fréchet derivative terms are missing. Of course, it is easy to modify the iterations in the boundary and side conditions to include these terms. However, this will cause technical difficulties, since the index of the boundary value problem will change for each t_j.

Now, in contrast to (P_t), the problems for w_m are all linear and hence are solvable in principle. Taking for granted the solvability of the linear problem (2.3), as for any iteration scheme, we still have to face the question of the convergence of the sequence defined by (2.3). For this purpose, let us consider the difference $\omega_m := w_{m+1}-w_m$. This will provide us information concerning what kind of regularity conditions for the coefficients and given data in the boundary value problem one should expect. It is clear from the definition of w_m that for $m = 0,1,\ldots,$ ω_m should be the solution of the linear boundary value problem:

$$(2.4)\quad \begin{aligned} &L\omega_m = \nu_{1m}\omega_m+\nu_{2m}\overline{\omega_m}+\lambda_m && \text{in} \quad D \\ &\operatorname{Re}\{e^{i\tau}\omega_m\} = \psi_m && \text{on} \quad \Gamma \\ &\Sigma^{-1}\int_\Gamma \operatorname{Im}\{\omega_m\}\sigma\, ds = \kappa_m \\ &\omega_m(z_k) = a_{mk}, \quad z_k \in D, \quad k = 1,2,\ldots,|n|, \end{aligned}$$

where

$$\nu_{1m} = t_j h_w(z,w_m), \quad \nu_{2m} = t_j h_{\bar{w}}(z,w_m), \quad \text{for} \quad m = 0,1,\ldots,$$

$$(2.5)_o\quad \begin{aligned} \lambda_o &= (t_j-t_{j-1})h(z,w_o), \\ \psi_o &= (t_j-t_{j-1})\psi(z,w_o), \\ \kappa_o &= (t_j-t_{j-1})\kappa(w_o), \\ a_{ok} &= (t_j-t_{j-1})a_k(w_o) \end{aligned}$$

and for $m \geq 1$,

$$\lambda_m = t_j\left\{\frac{1}{2}\, h_{(w,w)}\omega_{m-1}^2+h_{(w,\bar{w})}|\omega_{m-1}|^2+ \frac{1}{2}\, h_{(\bar{w},\bar{w})}\bar{\omega}_{m-1}^2\right\}$$

$$(2.5)_m\quad \begin{aligned} \psi_m &= t_j\{\psi(z,w_m)-\psi(z,w_{m-1})\} \\ \kappa_m &= t_j\{\kappa(w_m)-\kappa(w_{m-1})\} \\ a_{mk} &= t_j\{a_k(w_m)-a_k(w_{m-1})\}. \end{aligned}$$

Here we have tacitly applied the mean-value theorem to $h(z,w_m)$ and $h_{(w,w)}$ is defined by

$$h_{(w,w)} = \int_0^1 h_{ww}(z,\tau w_{m-1}+(1-\tau)w_m)\, d\tau,$$

and similarly for $h_{(w,\bar{w})}$ and $h_{(\bar{w},\bar{w})}$.

From these formulations, it is clear that we need some kind of a priori estimates for the linear problems such as (2.4) in order to estimate ω_m. Also from (2.5) this suggests that the t-step size chosen in the iteration procedure must satisfy certain criterion depending on bounds for the data. All of these cannot be established without the precise regularity conditions on the given coefficients and data in the problem which we shall discuss in the next section.

3. Main Results

In what follows, we denote by $C^{k+\alpha}(\hat{D})$ $(0 < \alpha \leq 1)$ the space of k-time Hölder continuously differentiable functions on $\hat{D}$ equipped with the usual Hölder norm $||\cdot||_{k+\alpha}$ and by $C^{k+\alpha}(\Gamma)$ the corresponding boundary function space with the norm $||\cdot||_{k+\alpha,\Gamma}$.

As we mentioned in the previous section the success of the method of embedding for treating nonlinear problems depends upon the suitable a priori estimates of the solutions to the corresponding linear problems. For our purpose, the following a priori estimate will be sufficient.

__Theorem 1.__ __Suppose that__ $\tau \in C^{1+\alpha}(\Gamma)$, $\sigma \in C(\Gamma)$ __and__ $q_1, q_2 \in C^{1+\alpha}(\hat{D})$ __satisfying the ellipticity condition__ (Q). __Then for any__ $w \in C^{1+\alpha}(\hat{D})$ __the estimate__

$$(3.1)\quad \begin{aligned} ||w||_{1+\alpha} \leq \gamma_1 ||\mathrm{Re}\,\{e^{i\tau}w\}||_{1+\alpha,\Gamma} &+ \gamma_2 \left|\Sigma^{-1}\int_\Gamma \mathrm{Im}\,\{w\}\sigma\, ds\right| \\ &+ \gamma_3 \sum_{k=1}^{|n|} |w(z_k)| + \gamma ||Lw - \nu_1 w - \nu_2 \overline{w}||_\alpha \end{aligned}$$

__holds for any__ $\nu_1, \nu_2 \in C^\alpha(\hat{D})$ __with__ $||\nu_1||_\alpha + ||\nu_2||_\alpha \leq K$. __The constants__ γ_1, γ_2, γ_3 __and__ γ __depend only on__ D, z_k $(1 \leq k \leq |n|)$, σ, τ, q_1, q_2 __and the constant__ K __but not on__ ν_1, ν_2 __or__ w.

The proof of Theorem 1 is lengthy and requires some technical details. In order not to interrupt the presentation, we will defer it to the next section.

In the following we shall establish an existence theorem based on the estimate in Theorem 1. To this end, we make the following assumptions:

(H-1) The complex-valued function $h(\cdot, w(\cdot))$ is defined in $\hat{D}\times\mathbb{C}$; for each fixed value of $w \in \mathbb{C}$, $h \in C^\alpha(\hat{D})$; h and its derivatives with respect to w and $\overline{w}$ up to the second order are Hölder-continuous with exponent α in $\hat{D}$; moreover there exists a constant K such that $||h||_\alpha \leq K$, $||h_w||_\alpha + ||h_{\overline{w}}||_\alpha \leq K$ and $||h_{ww}||_\alpha + 2||h_{w\overline{w}}||_\alpha + ||h_{\overline{w}\overline{w}}|| \leq 2K$.

(H-2) The real-valued function $\psi(\cdot, w(\cdot))$ is defined on $\Gamma\times\mathbb{C}$ such that $\psi(\cdot, w) \in C^{1+\alpha}(\Gamma)$ for all $w \in C^{1+\alpha}(\Gamma)$ and satisfies the Lipschitz condition:

$$||\psi(\cdot,w)-\psi(\cdot,\omega)||_{1+\alpha,\Gamma} \le L_1||w-\omega||_{1+\alpha,\Gamma} \quad \text{for all} \quad w,\omega \in C^{1+\alpha}(\Gamma),$$

where L_1 is a constant independent of w and ω.

(H-3) The real-valued functional κ satisfies the Lipschitz condition:

$$|\kappa(w)-\kappa(\omega)| \le L_2||w-\omega||_{1+\alpha,\Gamma} \quad \text{for all} \quad w,\omega \in C^{1+\alpha}(\Gamma),$$

where L_2 is a constant independent of w and ω.

(H-4) The complex-valued functional a_k satisfies the Lipschitz condition:

$$\sum_{k=1}^{|n|} |a_k(w)-a_k(\omega)| \le L_3||w-\omega||_{1+\alpha} \quad \text{for all} \quad w,\omega \in C^{1+\alpha}(\hat{D}),$$

where L_3 is a constant independent of w and ω.

In addition, we impose the restrictions on the coefficients:

(H-5) $\tau \in C^{1+\alpha}(\Gamma)$, $\sigma \in C(\Gamma)$ and $q_1,q_2 \in C^{1+\alpha}(\hat{D})$ satisfying (Q).

We now formulate the existence theorem:

<u>Theorem 2</u>. <u>Under the hypotheses</u> (H-1)-(H-5), <u>there exists a solution</u> $w \in C^{1+\alpha}(\hat{D})$ <u>to the semilinear Riemann-Hilbert boundary value problem</u>, (E), (B), (C) <u>and</u> (C') <u>which can be constructed by the method of imbedding, provided the Lipschitz constants</u> L_1, L_2 <u>and</u> L_3 <u>in the hypotheses are sufficiently small</u>.

We begin the proof of Theorem 2 with the existence theorem for the linear Riemann-Hilbert boundary value problem below.

<u>Lemma 3.1</u>. <u>For given</u> $\lambda \in C^{\alpha}(\hat{D})$, $f \in C^{1+\alpha}(\Gamma)$, $\chi \in \mathbb{R}$ <u>and</u> $c_k \in \mathbb{C}$, $k = 1,2,\ldots,|n|$, <u>there exists a unique solution</u> $w \in C^{1+\alpha}(\hat{D})$ <u>to the linear boundary value problem</u>:

$$(3.2) \qquad \begin{aligned} &Lw = \nu_1(z)w+\nu_2(z)\bar{w}+\lambda(z) && \underline{\text{in}} \ D \\ &\text{Re}\,\{e^{i\tau}w\} = f(z) && \underline{\text{on}} \ \Gamma \\ &\Sigma^{-1}\int_\Gamma \text{Im}\,\{e^{i\tau}w\}\sigma\, ds = \chi \\ &w(z_k) = c_k, \quad z_k \in D, \quad k = 1,2,\ldots,|n| \end{aligned}$$

<u>with</u> $\nu_1,\nu_2 \in C^{\alpha}(\hat{D})$ <u>and under the assumption</u> (H-5).

Proof. This lemma holds for the special case when $q_1 = q_2 = 0$ [17, Theorem 1.2.4]. The general case can be reduced to the special case by some elementary transformations (see Lemma 4.2) and the inverse transformations lead to a solution of (3.2). The uniqueness follows at once from the a priori estimate (3.1).

Based on Lemma 3.1, it is clear that under the assumptions (H-1)-(H-5), the linear problem (2.3) for w_{m+1} is uniquely solvable and the solution w_{m+1} is again in $C^{1+\alpha}(\hat{D})$ if $w_m \in C^{1+\alpha}(\hat{D})$. Now let us assume $w(z,t_{j-1}) \in C^{1+\alpha}(\hat{D})$ to be the solution of (P_t) for a fixed t_{j-1}, $0 \le t_{j-1} < 1$, and define the sequence $\{w_m\}$ according to (2.2) and (2.3) for $t = t_j$, $t_{j-1} < t_j \le 1$. It remains now to consider the difference $\omega_m = w_{m+1}-w_m$ in (2.4).

From the a priori estimate (3.1) together with the regularity assumptions (H-1)-(H-4), we obtain

$$
\begin{aligned}
||\omega_o||_{1+\alpha} &\le (t_j-t_{j-1})\{\gamma_1||\psi(\cdot,w_0)||_{1+\alpha,\Gamma}+\gamma_2|\kappa(w_o)| \\
&\quad +\gamma_3\sum_{k=1}^{|n|}|a_k(w_0)|+\gamma||h(\cdot,w_o)||_\alpha\} \qquad (3.3)\\
&\le (t_j-t_{j-1})\{\beta||w_0||_{1+\alpha}+\beta_o\}
\end{aligned}
$$

where

$$\beta := \gamma_1L_1+\gamma_2L_2+\gamma_3L_3 \qquad (3.4)$$

and

$$\beta_o := \gamma_1||\psi(\cdot,0)||_{1+\alpha,\Gamma}+\gamma_2|\kappa(0)|+\gamma_3\sum_{k=1}^{|n|}|a_k(0)|+\delta \qquad (3.5)$$

with $\delta = \kappa\gamma$. Similarly, we have for $m \ge 1$

$$
\begin{aligned}
||\omega_m||_{1+\alpha} &\le t_j\{\beta||\omega_{m-1}||_{1+\alpha,\Gamma}+\delta||\omega_{m-1}||_\alpha^2\} \qquad (3.6)\\
&\le t_j\{\beta+\hat{\delta}||\omega_{m-1}||_{1+\alpha}\}||\omega_{m-1}||_{1+\alpha}.
\end{aligned}
$$

The rest of the proof follows our previous papers [2], [3], exactly. In particular, if we require $\beta < 1$, it is not difficult to see that the sequence $\{w_m\}$ will converge to $w(z,t_j) \in C^{1+\alpha}(\hat{D})$, provided that t_j is chosen near t_{j-1} enough so that

$$t_j-t_{j-1} < \frac{(1-\beta)^2}{\delta\beta_o}. \qquad (3.7)$$

Moreover, (3.7) gives an upper bound for t_j-t_{j-1} which is independent of t_{j-1}. This implies that after finitely many steps one can eventually reach to the limit $w(z,1)$ which is a desired solution of the boundary value problem, (E), (B), (C) and (C'). This completes the proof of Theorem 2.

As a consequence of the a priori estimate (3.1), the following uniqueness result can easily be established:

Theorem 3. *Under the same assumptions of Theorem 2, the semi-linear Riemann-Hilbert boundary value problem*, (E), (B), (C) *and* (C') *is uniquely solvable in* $C^{1+\alpha}(\hat{D})$.

In practice, one may use the solution $w_{m+1}(z,t_j)$ of (2.3) for some fixed t_j and m to approximate the exact solution $w(z) = w(z,1)$. The following error estimate provides us information concerning the rate of convergence as well as its accuracy.

Theorem 4. *Let* γ_o *be any given real number such that* $\beta < \gamma_o < 1$. *Then for any fixed* t_{j-1}, $0 \le t_{j-1} < 1$, *if we choose* t_j *satisfying* $1 \ge t_j > t_{j-1} \ge 0$ *and*

$$t_j - t_{j-1} \le \frac{(\gamma_o-\beta)(1-\beta)}{\beta_o\delta} \quad \text{for} \quad j = 1,2,\ldots$$

the error estimate

$$||w-w_{m+1}(\cdot,t_j)||_{1+\alpha} \le \frac{\beta_o}{(1-\beta)^2}(1-t_j) + \frac{\gamma_o-\beta}{\delta(1-\beta)}(t_j\gamma_o)^{m+1}$$

holds, where β *and* β_o *are defined by* (3.4) *and* (3.5) *respectively*.

This estimate again relies on the a priori estimate (3.1) and can be obtained in a similar manner as for ω_m in Theorem 2, if one splits the estimate according to the form:

$$w(z)-w_{m+1}(z,t_j) = [w(z,1)-w(z,t_j)]+[w(z,t_j)-w_{m+1}(z,t_j)].$$

We omit the details.

4. The Derivation of A-Priori Estimates

It is clear by now the a priori estimate (3.1) is crucial to the applicability of the embedding method. In this section we shall establish the result (3.1). The derivation is rather long and involved. We shall only sketch the proof and leave out some of the details which will be available in our forthcoming paper [4]. With a slight change of the conditions in Theorem 1, we may even have a more general result below:

Theorem 5. *Suppose that* $\tau \in C^{1+\alpha}(\Gamma)$ *and* $\sigma \in C(\Gamma)$. *Then for any* $w \in C^{1+\alpha}(\hat{D})$, *there are constants* $\gamma_1, \gamma_2, \gamma_3$ *and* γ *such that*

$$(4.1)\qquad \begin{aligned} ||w||_{1+\alpha} &\le \gamma_1 ||\operatorname{Re}\{e^{i\tau}w\}||_{1+\alpha,\Gamma} + \gamma_2 \left|\Sigma^{-1}\int_\Gamma \operatorname{Im}\{w\}\sigma\, ds\right| \\ &+ \gamma_3 \sum_{k=1}^{|n|} |w(z_k)| + \gamma ||w_{\bar z} - \mu_1 \bar w_z - \mu_2 w_z - \nu_1 w - \nu_2 \bar w||_\alpha \end{aligned}$$

for any $\mu_1, \mu_2 \in C^{1+\alpha}(\mathbb{C})$ *and* $\nu_1, \nu_2 \in C^\alpha(\hat{D})$ *with the properties*:

(a) $\mu_1 = \mu_2 = 0$ *in* $\mathbb{C}\setminus\hat{D}$; $||\mu_1||_o + ||\mu_2||_o \le q < 1$,

$||\mu_{1z}||_\alpha + ||\mu_{2z}||_\alpha + ||\mu_{1\bar z}||_\alpha + ||\mu_{2\bar z}||_\alpha \le M_1$ *and*

(b) $||\nu_1||_\alpha + ||\nu_2||_\alpha \le K$.

The constants $\gamma_1, \gamma_2, \gamma_3$ *and* γ *depend only on* D, z_k $(1 \le k \le |n|)$, σ, τ, α *and the constants* q, M_1, K *but not on* μ_1, μ_2, ν_1, ν_2 *or* w.

We remark that if μ_1 and μ_2 are only defined on $\hat{D}$ such as q_1 and q_2 in Theorem 1, one may always extend them by using the standard cut-off functions. The derivation of (4.1) will be proceeded according to the outline: First, we establish the result (4.1) for the special case when μ_1 and μ_2 are identically zero. Next by using elementary transformations we show that the general case can be reduced to the special case. Finally, estimates are obtained by using the available result for the special case.

To this end we state without proofs the following lemmas. The proofs are given in [4].

Lemma 4.1. *Let* $\tau \in C^{1+\alpha}(\Gamma)$ *and* $\sigma \in C(\Gamma)$ *be given. Then for any* $w \in C^{1+\alpha}(\hat{D})$ *there are constants*, $\tilde\gamma_1, \tilde\gamma_2, \tilde\gamma_3$ *and* $\tilde\gamma$ *such that*

$$(4.2)\qquad \begin{aligned} ||w||_{1+\alpha} &\le \tilde\gamma_1 ||\operatorname{Re}\{e^{i\tau}w\}||_{1+\alpha,\Gamma} + \tilde\gamma_2 \left|\Sigma^{-1}\int_\Gamma \operatorname{Im}\{w\}\sigma\, ds\right| \\ &+ \tilde\gamma_3 \sum_{k=1}^{|n|} |w(z_k)| + \tilde\gamma ||w_{\bar z} - aw - b\bar w||_\alpha \end{aligned}$$

holds for any $a, b \in C^\alpha(\hat{D})$ *with* $||a||_\alpha + ||b||_\alpha \le K$, *where the constants depend only on* D, z_k $(1 \le k \le |n|)$, τ, σ, α *and the constant* K, *but not on* a, b *or* w.

It should be mentioned that an estimate similar to (4.2) based on the closed graph theorem is also available in [17,

Eq. (1.2.31)]. However, we emphasize that in contrast to [17], the constants $\tilde{\gamma}_1$, $\tilde{\gamma}_2$, $\tilde{\gamma}_3$ and $\tilde{\gamma}$ appearing in (4.2) depend not on a and b but only on the boundedness of a and b. Our derivation of (4.2) in [4] follows [2] and [3] by using Green's formula and Privaloff's theorem.

In order to state the next lemma which concerns the transformation, we need the following. For $\mu_1, \mu_2 \in C^{1+\alpha}(\mathbb{C})$ satisfying $|\mu_1|+|\mu_2| \leq q < 1$, we define

$$(4.3) \qquad \mu := \frac{2\mu_1}{1+|\mu_1|^2-|\mu_2|^2-\sqrt{\Delta}} \quad \text{and} \quad a := \frac{2\mu_2}{1-|\mu_1|^2+|\mu_2|^2-\sqrt{\Delta}}$$

where $\Delta = (1-|\mu_1|^2-|\mu_2|^2)^2-4|\mu_1\mu_2|^2$. We denote by ζ the complete homeomorphism of the Beltrami system [12]

$$\zeta_{\bar{z}} = \mu\zeta_z,$$

where μ is defined by (4.3); and by $\tilde{D}$ and $\tilde{\Gamma}$ respectively the images of D and Γ under the homeomorphism ζ. We remark that μ and a are well-defined, since $|\mu_1|+|\mu_2| < 1$ and $\Delta \geq [1-(|\mu_1|+|\mu_2|)^2]^2$. Moreover, it can be shown that both $|\mu|$ and $|a|$ are dominated by $|\mu_1|+|\mu_2|$.

Lemma 4.2. *Suppose that the coefficients* μ_1, μ_2, ν_1, ν_2, τ *and* σ *satisfy the assumptions in Theorem 5. In addition, assume that* $\lambda \in C^\alpha(\hat{D})$, $f \in C^{1+\alpha}(\Gamma)$, $\chi \in R$ *and* $c_k \in \mathbb{C}$, $k = 1,\ldots,|n|$. *Then the linear boundary value problems defined by*

$$(4.4) \quad \begin{cases} w_{\bar{z}} = \mu_1(z)w_z+\mu_2(z)\bar{w}_z+\nu_1(z)w+\nu_2(z)\bar{w}+\lambda(z) & \underline{\text{in}} \quad D \\ \operatorname{Re}\{e^{i\tau}w\} = f(z) & \underline{\text{on}} \quad \Gamma \\ \Sigma^{-1}\int_\Gamma \operatorname{Im}\{e^{i\tau}w\}\sigma\, ds = \chi \\ w(z_k) = c_k, \quad z_k \in D, \quad k = 1,2,\ldots,|n| \end{cases}$$

and

$$(4.5) \quad \begin{cases} \tilde{\omega}_{\bar{\zeta}} = A(\zeta)\tilde{\omega}+B(\zeta)\bar{\tilde{\omega}}+C(\zeta) & \underline{\text{in}} \quad \tilde{D} \\ \operatorname{Re}\{e^{i\tilde{\tau}}\tilde{\omega}\} = \tilde{f}(\zeta) & \underline{\text{on}} \quad \tilde{\Gamma} \\ \tilde{\Sigma}^{-1}\int_{\tilde{\Gamma}} \operatorname{Im}\{e^{i\tilde{\tau}}\tilde{\omega}\}\tilde{\sigma}\, d\tilde{s} = \tilde{\chi} \\ \tilde{\omega}(\zeta_k) = \tilde{c}_k, \quad \zeta_k = \zeta(z_k) \in \tilde{D}, \quad k = 1,2,\ldots,|n| \end{cases}$$

are equivalent for $w \in C^{1+\alpha}(\hat{D})$ *and* $\tilde{\omega} \in C^{1+\alpha}(\hat{\tilde{D}})$ *under the* transformation:

(4.6) $$\zeta_{\bar z} = \mu\zeta_z \quad \text{and} \quad \tilde\omega = \frac{\rho e^{i\phi}}{1-|a|^2}(w-a\bar w),$$

where μ and a are defined by (4.3); ρ and ϕ are the interior harmonic extensions of the real-valued functions ρ and $\phi \in C^{1+\alpha}(\Gamma)$ such that

(4.7) $$\rho e^{i\phi} = 1+\bar a e^{-2i\tau} \quad \text{on} \quad \Gamma.$$

The coefficients and data in (4.5) are related to those in (4.4) by the following expressions explicitly:

$$A := \frac{(1-\bar\mu_1\mu)(\nu_1+\bar a\nu_2)+\mu_2\mu(\bar a\bar\nu_1+\bar\nu_2)}{(1-|a|^2)(|1-\bar\mu_1\mu|^2-|\mu_2\mu|^2)\bar\zeta_z} + \frac{\bar a_\zeta a}{1-|a|^2} + \frac{\rho_{\bar\zeta}}{\rho} + i\phi_{\bar\zeta},$$

$$B := \left[\frac{(1-\bar\mu_1\mu)(a\nu_1+\nu_2)+\mu_2\mu(\bar\nu_1+a\bar\nu_2)}{(1-|a|^2)(|1-\bar\mu_1\mu|^2-|\mu_2\mu|^2)\bar\zeta_z} - \frac{a_{\bar\zeta}}{1-|a|^2}\right]e^{2i\phi},$$

$$C := \left(\frac{(1-\bar\mu_1\mu)\lambda+\mu_2\mu\bar\lambda}{(|1-\bar\mu_1\mu|^2-|\mu_2\mu|^2)\bar\zeta_z}\right)\frac{\rho e^{i\phi}}{1-|a|^2},$$

(4.8)
$$\tilde f(\zeta) := f(z(\zeta)), \qquad \tilde\tau(\zeta) := \tau(z(\zeta)),$$

$$\tilde\chi := \Sigma\tilde\Sigma^{-1}\chi - \tilde\Sigma^{-1}\int_\Gamma \frac{2\,\mathrm{Im}\,(ae^{i\tau})}{|1+ae^{2i\tau}|^2} f\sigma\, ds,$$

$$\tilde c_k := \frac{\rho(z_k)e^{i\phi(z_k)}}{1-|a(z_k)|^2}(c_k+a(z_k)\bar c_k);$$

moreover,

(4.9) $$\tilde\sigma := \frac{1-|a|^2}{|1+ae^{2i\tau}|^2}\frac{ds}{d\tilde s}\sigma \quad \text{and} \quad \tilde\Sigma = \int_{\tilde\Gamma}\tilde\sigma\, d\tilde s.$$

With these two lemmas available, we now turn our attention to the estimate (4.1). From the definition of A, B and C, it can be shown there exist constants $\tilde K$ and M such that

$$||A||_{\alpha,\hat{\tilde D}}+||B||_{\alpha,\hat{\tilde D}} \le \tilde K \quad \text{and} \quad ||C||_{\alpha,\hat{\tilde D}} \le M||\lambda||_{\alpha,\hat D}.$$

Then Lemma 4.1 ensures the existence of constants $\tilde\gamma_1$, $\tilde\gamma_2$, $\tilde\gamma_3$ and $\tilde\gamma$ depending on $\tilde D$, ζ_k $(1 \le k \le |n|)$, $\tilde\sigma$, $\tilde\tau$, α, $\tilde K$ or therefore on D, z_k $(1 \le k \le |n|)$, σ, τ, α, K, and M, such that

$$(4.10)\quad ||\tilde{\omega}||_{1+\alpha,\hat{\tilde{D}}} \le \tilde{\gamma}_1||\tilde{f}||_{1+\alpha,\tilde{\Gamma}}+\tilde{\gamma}_2|\tilde{\chi}|+\tilde{\gamma}_3\sum_{k=1}^{|n|}|\tilde{c}_k|+\tilde{\gamma}M||\lambda||_{\alpha,\hat{D}}.$$

Now from (4.6) it follows that

$$w = \frac{\tilde{\omega}}{\rho e^{i\phi}} + \frac{a\bar{\tilde{\omega}}}{\rho e^{-i\phi}}$$

and hence

$$(4.11)\quad \begin{aligned} ||w||_{1+\alpha,\hat{D}} &\le \{||\rho^{-1}e^{-i\phi}||_{1+\alpha,\hat{D}}+||a\rho^{-1}e^{i\phi}||_{1+\alpha,\hat{D}}\}||\tilde{\omega}||_{1+\alpha,\hat{D}} \\ &\le M_0||\tilde{\omega}||_{1+\alpha,\hat{D}} \end{aligned}$$

for some constant M_0. It can also be shown that

$$(4.12)\quad \begin{aligned} ||\tilde{\omega}_z||_{\alpha,\hat{D}}+||\tilde{\omega}_{\bar{z}}||_{\alpha,\hat{D}} \le Me^{\alpha MM_1}(1+q)^{\alpha}\{&||\tilde{\omega}_{\zeta}||_{\alpha,\hat{\hat{D}}} \\ &+||\tilde{\omega}_{\bar{\zeta}}||_{\alpha,\hat{\hat{D}}}\} \end{aligned}$$

by making use of the inequality

$$|\zeta_1-\zeta_2| \le e^{MM_1}(1+q)|z_1-z_2|.$$

Therefore, (4.12) implies

$$||\tilde{\omega}||_{1+\alpha,\hat{D}} \le Me^{\alpha MM_1}(1+q)^{\alpha}||\tilde{\omega}||_{1+\alpha,\hat{\hat{D}}}$$

from which we have, in view of (4.11),

$$(4.13)\quad ||w||_{1+\alpha,\hat{D}} \le \tilde{M}||\tilde{\omega}||_{1+\alpha,\hat{\hat{D}}}$$

for some constant $\tilde{M}$.

Similarly from the definitions (4.8), it is not difficult to show that

$$(4.14)\quad \begin{aligned} ||\tilde{f}||_{1+\alpha,\tilde{\Gamma}} &\le \left|\left|\frac{ds}{d\tilde{s}}\right|\right|_{\alpha,\tilde{\Gamma}}\frac{e^{\alpha MM_1}}{(1-q)^{\alpha}}||f||_{1+\alpha,\Gamma}, \\ |\tilde{\chi}| &\le \frac{1+q}{1-q}\{|\chi|+\frac{2q}{(1-q)^2}||f||_{0,\Gamma}\} \\ \text{and}\quad |\tilde{c}_k| &\le \frac{1+q}{1-q}|c_k|. \end{aligned}$$

Now collecting the results (4.10), (4.13), and (4.14), we arrive finally at the desired estimate (4.1) in view of the definitions of λ, f, χ and c_k in (4.4).

To conclude this paper, we would like to point out that with an a priori estimate such as (4.1) one may even be able to treat boundary value problems for quasilinear elliptic systems by the embedding method. Furthermore, with some modifications perhaps the assumptions on the smallness of the Lipschitz constants can be relaxed. Indeed, we will pursue these investigations in a separate communication.

References

[1] H. Begehr and R. P. Gilbert, Das Randwert-Normproblem für ein fastlineares elliptisches System und eine Anwendung, Ann. Acad. Sci. Fenn, AI, 3 (1977), 179-184.

[2] H. Begehr and G. C. Hsiao, On nonlinear boundary value problems of elliptic systems in the plane, Lecture Notes in Math., No. 846, Springer-Verlag (1981), 55-63.

[3] H. Begehr and G. C. Hsiao, Nonlinear boundary value problems for a class of elliptic systems, in Komplex Analysis und ihre Anwendung anf partielle Differentialgleichungen, Martin-Luther-Universität, Halle-Wittenberg (1980), 90-102.

[4] H. Begehr and G. C. Hsiao, A priori estimates for elliptic systems, to appear.

[5] H. Begehr and G. N. Hile, Nonlinear Riemann boundary value problems for a semilinear elliptic system in the plane, to appear.

[6] B. Bojarski, Quasiconformal mappings and general structural properties of systems of non linear equations elliptic in the sense of Lavrent'ev, Inst. Naz. Alta Mat. Symp. Math. 18 (1976), 485-499.

[7] B. Bojarski, Subsonic flow of compressible fluid, Archiwum Mechaniki Stosowanej 4, 18 (1966), 497-520.

[8] R. P. Gilbert, Nonlinear boundary value problems for elliptic systems in the plane, Proc. Intl. Conf. Nonlinear Systems Appl., ed. V. Lakshmikantham, 1977, Academic Press, 97-124.

[9] T. V. Iwaniec, Quasiconformal mapping problem for general nonlinear systems of partial differential equations, Inst. Naz. Alta Mat. Symp. Math. 18 (1976), 501-517.

[10] E. V. Tjurikov, The nonlinear Riemann-Hilbert boundary value problem for quasilinear elliptic systems, Soviet Math. Dokl. 20 (1979), 863-866.

[11] W. Tutschke, The Riemann-Hilbert problem for nonlinear systems of differential equations in the plane, Complex analysis and its applications, Akad. Nauk SSSR, Moscow, 1978, 537-542 (Russian).

[12] I. N. Vekua, Generalized Analytic Functions, Pergamon Press, 1962.

[13] H. J. Wacker, Eine Lösungsmethode zur Behandlung nichtlinearer Randwertprobleme, Internationsverfahren, Numerische Mathematik, Approximationstheorie ISNM, V. 15, Birkhauser (1970), 245-257.

[14] G. Warowna-Dorau, Application of the method of successive approximations to a non-linear Hilbert problem in the class of generalized analytic functions, Demonstratio Math., 2 (1970), 101-116.

[15] W. Wendland, An integral equation method for generalized analytic functions, Lecture Notes in Math., No. 430, Springer-Verlag (1974), 414-452.

[16] W. Wendland, On the imbedding method for semilinear first order elliptic systems and related finite element methods, in Continuation Methods, ed. H. Wacker, Academic Press, 1977, 277-336.

[17] W. Wendland, Elliptic Systems in the Plane, Pitman Publishing, Inc., London, 1978.

[18] J. Wolska-Bochenek, A compound non-linear boundary value problem in the theory of pseudo-analytic functions, Demonstratio Math., 4 (1972), 105-117.

Contemporary Mathematics
Volume 11, 1982

Spinor valued regular functions

by

Pertti Lounesto
Institute of Mathematics
Helsinki University of Technology
02150 Espoo 15, Finland

Abstract Clifford algebras and spinors appear in various physical applications, e.g., the Maxwell equations can be condenced into one equation employing the Clifford notation, the Dirac equation employes Dirac spinors, and the Kepler motion can be regularized employing the spinor regularization. Therefore, it seems desirable to develop a theory of functions with values in the Clifford algebra. In this paper, an algebraic back-ground of the spinor function theory is established. The authors consider differentiable functions from the euclidean space R^n to some minimal left ideal V of the Clifford algebra R_n on R^n. Such a function is called regular if it satisfies a higher dimensional analogue of the Cauchy-Riemann equations, and these regular functions satisfy, e.g., a higher dimensional analogue of Cauchy's integral formula. Since the domain R^n and the target V have norms invariant under the rotation group $SO(n)$ of R^n, all those theorems of classical function theory, which depend only on Cauchy's integral formula and on the existence of a norm, e.g., Liouville's theorem and maximum modulus principle, have natural generalizations in higher dimensions.

0271-4132/82/0000-0147/$06.25

1. Cauchy-Riemann equations in Clifford notation

The Clifford algebra R_2 on the euclidean plane R^2 is a four-dimensional real associative algebra generated by an orthonormal basis (e_1, e_2) for R^2 with the relations

$$e_1^2 = e_2^2 = 1, \quad e_2 e_1 = -e_1 e_2 .$$

As a real linear space the Clifford algebra R_2 is spanned by the elements $1, e_1, e_2, e_{12}$ where $e_{12} = e_1 e_2$. The following multiplication table of the Clifford algebra R_2 can be verified by direct computation

1	e_1	e_2	e_{12}
e_1	1	e_{12}	e_2
e_2	$-e_{12}$	1	$-e_1$
e_{12}	$-e_2$	e_1	-1

The correspondences

$$1 \simeq \begin{pmatrix} 1 & 0 \\ 0 & 1 \end{pmatrix}, \; e_1 \simeq \begin{pmatrix} 1 & 0 \\ 0 & -1 \end{pmatrix}, \; e_2 \simeq \begin{pmatrix} 0 & 1 \\ 1 & 0 \end{pmatrix}, \; e_{12} \simeq \begin{pmatrix} 0 & 1 \\ -1 & 0 \end{pmatrix}$$

establish an isomorphism $R_2 \simeq R(2)$ between the Clifford algebra R_2 and the algebra of real 2×2-matrices $R(2)$.

Consider next a region G in the euclidean plane R^2 and a continuously differentiable function $G \to R^2$ sending a vector $xe_1 + ye_2$ to a vector $ue_1 + ve_2$. Such a function

$ue_1 + ve_2$ will be called regular in G if the condition

$$(e_1 \frac{\partial}{\partial x} + e_2 \frac{\partial}{\partial y}) (ue_1 + ve_2) = 0$$

holds for all $xe_1 + ye_2$ in $\dot{G}$. This condition gives us indirectly conformal maps of the euclidean plane R^2 unless the derivative of the map is zero, cf. [11].

In order to obtain directly conformal maps of the euclidean plane R^2 we decompose the operator

$$(xe_1 + ye_2)(e_1 \frac{\partial}{\partial x} + e_2 \frac{\partial}{\partial y}) = (x \frac{\partial}{\partial x} + y \frac{\partial}{\partial y})$$

$$+ e_{12}(x \frac{\partial}{\partial y} - y \frac{\partial}{\partial x})$$

into its homogeneous components of degree 0 and 2. In the polar coordinates $xe_1 + ye_2 = r(\cos \varphi\, e_1 + \sin \varphi\, e_2)$ these homogeneous components

$$r \frac{\partial}{\partial r} = x \frac{\partial}{\partial x} + y \frac{\partial}{\partial y}, \quad e_{12} \frac{\partial}{\partial \varphi} = e_{12}(x \frac{\partial}{\partial y} - y \frac{\partial}{\partial x})$$

are seen to contain only radial and angular operations, respectively. Moreover, $e_1 e_2 = e_r e_\varphi$ where $e_r = \cos \varphi\, e_1 + \sin \varphi\, e_2$ and $e_\varphi = -\sin \varphi\, e_1 + \cos \varphi\, e_2$. Clearly, the function $ue_1 + ve_2$ is regular in $G \setminus 0$ if the condition

$$(r \frac{\partial}{\partial r} + e_{12} \frac{\partial}{\partial \varphi})(ue_1 + ve_2) = 0$$

is satisfied in G. The function $ue_1 + ve_2$ will be called antiregular in $G \setminus 0$ if

$$(r \frac{\partial}{\partial r} - e_{12} \frac{\partial}{\partial \varphi})(ue_1 + ve_2) = 0$$

holds in G. It follows that the antiregular vector valued functions satisfy the Cauchy-Riemann equations.

The vector space R^2 is itself an irreducible representation module of the orthogonal group $O(2)$ of R^2. In addition, any minimal left ideal V of R_2 is a two-dimensional irreducible representation module of $Pin(2)$. The element $f = \frac{1}{2}(1 + e_1)$ is a primitive idempotent of the Clifford algebra R_2, and the subspace $V = R_2 f$ of R_2, spanned by the elements

$$f_1 = e_1 f \simeq \begin{pmatrix} 1 & 0 \\ 0 & 0 \end{pmatrix}, \quad f_2 = e_2 f \simeq \begin{pmatrix} 0 & 0 \\ 1 & 0 \end{pmatrix},$$

is a minimal left ideal of R_2. The elements of this module V of $Pin(2)$ are called spinors.

Consider a continuously differentiable function $R^2 \supset G \to V$ sending a vector

$$xe_1 + ye_2 \simeq \begin{pmatrix} x & y \\ y & -x \end{pmatrix}$$

to a spinor

$$uf_1 + vf_2 \simeq \begin{pmatrix} u & 0 \\ v & 0 \end{pmatrix}.$$

Such a spinor valued function $uf_1 + vf_2$ will be called <u>regular</u> in G if

$$\left(e_1 \frac{\partial}{\partial x} + e_2 \frac{\partial}{\partial y}\right)(uf_1 + vf_2) = 0$$

holds in G, and <u>antiregular</u> in $G \setminus O$ if

$$\left(r \frac{\partial}{\partial r} - e_{12} \frac{\partial}{\partial \varphi}\right)(uf_1 + vf_2) = 0$$

holds in G. Since

$$e_1 f_1 = f_1 \qquad e_1 f_2 = -f_2$$
$$e_2 f_1 = f_2 \qquad e_2 f_2 = f_1$$

one verifies that these equations impose the same conditions on the functions u and v as the corresponding earlier equations.

Before closing this chapter we emphasize that the differential operators

$$P = e_1 \frac{\partial}{\partial x} + e_2 \frac{\partial}{\partial y}$$

$$D = x \frac{\partial}{\partial x} + y \frac{\partial}{\partial y}$$

$$M = e_{12}\left(x \frac{\partial}{\partial y} - y \frac{\partial}{\partial x}\right)$$

are invariant under the orthogonal group $O(2)$ of the euclidean plane R^2.

2. Invariant differential operators

Let us consider an n-dimensional euclidean space R^n, and an orthonormal basis $(e_1, e_2, \ldots, e_n)$ for R^n. We shall need the Clifford algebra R_n on R^n, that is, the real associative algebra of dimension 2^n generated by R^n with the relations

$$e_\mu e_\nu + e_\nu e_\mu = 2\delta_{\mu\nu} .$$

Denote $e_{\mu\nu} = e_\mu e_\nu$ in short. The Clifford algebras R_n, $n < 8$, are isomorphic to the following matrix algebras

n	R_n
0	R
1	${}^2R$
2	$R(2)$
3	$C(2)$
4	$H(2)$
5	${}^2H(2)$
6	$H(4)$
7	$C(8)$

and this table continues with a periodicity of 8, cf. [6]. This expresses the periodicity theorem of Bott [1] in disguise.

In addition we shall need the differential operators generating rotations, translations, dilatations, and special conformal transformations

$$M_{\mu\nu} = x_\mu \frac{\partial}{\partial x_\nu} - x_\nu \frac{\partial}{\partial x_\mu}$$

$$P_\mu = \frac{\partial}{\partial x_\mu}$$

$$D = \sum_{\nu=1}^{n} x_\nu \frac{\partial}{\partial x_\nu}$$

$$K_\mu = \sum_{\nu=1}^{n} \left((x_\nu x_\nu) \frac{\partial}{\partial x_\mu} - 2x_\mu \left(x_\nu \frac{\partial}{\partial x_\nu}\right)\right)$$

of the euclidean space R^n. These differential operators form a basis for the Lie algebra of the conformal group of R^n, and they satisfy the commutation relations

$$[M_{\alpha\beta}, M_{\mu\nu}] = \delta_{\alpha\nu}M_{\beta\mu} - \delta_{\alpha\mu}M_{\beta\nu} + \delta_{\beta\mu}M_{\alpha\nu} - \delta_{\beta\nu}M_{\alpha\mu}$$

$$[P_\lambda, M_{\mu\nu}] = \delta_{\lambda\mu}P_\nu - \delta_{\lambda\nu}P_\mu \ , \ [D, P_\mu] = -P_\mu$$

$$[K_\lambda, M_{\mu\nu}] = \delta_{\lambda\mu}K_\nu - \delta_{\lambda\nu}K_\mu \ , \ [D, K_\mu] = K_\mu$$

$$[K_\mu, P_\nu] = 2(\delta_{\mu\nu} D + M_{\mu\nu}) \ ,$$

cf. [10]. Next, define

$$M = \sum_{\mu<\nu} e_{\mu\nu} M_{\mu\nu}$$

$$P = \sum_{\mu=1}^{n} e_\mu P_\mu$$

$$D = \sum_{\nu=1}^{n} x_\nu P_\nu$$

$$K = \sum_{\mu=1}^{n} e_\mu K_\mu \ .$$

These differential operators are invariant under the orthogonal group $O(n)$ of R^n. We shall employ them to define regularity and antiregularity for differentiable functions with values in the Clifford algebra R_n on R^n. Let now G be a region of R^n. A real differentiable function $G \to R_n$, $x \to f(x)$ will be called <u>regular</u> in G if

(2.1) $Pf(x) = 0$

for all x in G. Decompose the product xP into its homogeneous components

$$xP = D + M$$

of degree 0 and 2, respectively. Then

$$(D + M)f(x) = 0$$

implies $Pf(x) = 0$ when $x \neq 0$. A real differentiable function $G \to R_n$, $x \to f(x)$ will be called <u>antiregular</u> in $G \setminus 0$ if

$$(D - M)f(x) = 0$$

for all x in G. Since $K = x^2P - 2xD$, and therefore $K = -x(D - M)$, a function f is antiregular in $G \setminus 0$ if and only if

(2.2) $Kf(x) = 0$

for all x in G.

These differential operators satisfy, for example, the anticommutation relations

$$MP + PM = (n-1)P$$
$$MK + KM = (n-1)K$$
$$Mx + xM = (n-1)x$$
$$Px + xP = n+2D$$

which may be verified by direct computation.

<u>2.3 Example</u>. Since $Px = n$, $Dx = x$, and $Mx = (n-1)x$, we find that the identity map of R^n is not regular for any positive integer n, and it is antiregular only for $n = 2$. □

The following example shows that there exist non-trivial special regular functions in higher dimensions. Denote $x^{-1} = x(1/x^2)$, for $x \neq 0$.

<u>2.4 Example</u>. In the n-dimensional euclidean space R^n the function $q(x) = x|x|^{-n}$ is regular for $x \neq 0$, and the function $q(x^{-1}) = x^{-1}|x|^n$ is antiregular for $x \neq 0$. □

A differentiable function $f: G \to R^n$ is regular in G if and only if it is the gradient of a harmonic function $\emptyset$ in G, that is, $f = P\emptyset$ and $P^2\emptyset = 0$. For similar reasons it is not worth studying the special regular vector valued functions in R^n.

As a plausible higher dimensional analogue of the Cauchy-Riemann equations we shall consider antiregular spinor valued functions. Spinors are elements of a minimal left ideal V of R_n, the real dimension of V being approximately $2^{n/2}$, or for $n < 8$ more precisely

n	$\dim_R V$
0	1
1	1
2	2
3	4
4	8
5	8
6	16
7	16

and this table continues with a periodicity of 8.

<u>2.5 Spin and orbital angular momentum</u>. In the non-relativistic quantum mechanics of the electron spin the wave functions are spinor valued eigenstates of the differential bivector operator $M = x \wedge P$ in case $n = 3$; an analogous operator exists in any dimension $n \neq 3$, that is,

$$M\psi_\kappa = \kappa\psi_\kappa \; .$$

In general we have in R^n

$$M^2 = -\sum_{\mu<\nu} M_{\mu\nu}^2 + (n-2)M$$

because of the identity (for a Casimir operator)

$$M_{\alpha\lambda}M_{\mu\nu} + M_{\alpha\mu}M_{\nu\lambda} + M_{\alpha\nu}M_{\lambda\mu} = 0 .$$

For the value ℓ of the orbital angular momentum

$$\sum_{\mu<\nu} M_{\mu\nu}^2 \psi_\kappa + \ell(\ell + n - 2)\psi_\kappa = 0$$

and

$$\kappa^2 = \ell(\ell + n - 2) + (n - 2)\kappa .$$

Therefore

$$\kappa = \begin{cases} \ell + n - 2 & \text{spin up} \\ -\ell \quad (\ell \neq 0) & \text{spin down} \end{cases}$$

corresponding to the fact that the tensor product of the spin representation and the orbital angular momentum representation reduces in the spin-orbit coupling. □

The above differential operators appear in various physical applications. For example, the Maxwell equations can be condensed into one equation PF = j where F is the electromagnetic bivector field and j is the current vector field which can be considered as a given source, cf. [11]. In the Dirac equation P operates on Dirac spinors. Spinors appear also directly in some applications, for example, in the spinor regularization of the Kepler motion introduced by Kustaanheimo [7], see also [8].

3. Stokes' theorem and Cauchy's integral formula

In the following we shall consider in R^n integration of functions with values in the Clifford algebra R_n. The integration will be taken over a p-dimensional differentiable and oriented manifold S with boundary ∂S. The oriented k-volume of a small parallelepiped in S with $x_1, x_2, \ldots, x_k$ as adjacent edges is the k-vector part of the product $x_1 x_2 \ldots x_k$, that is, the antisymmetric product

$$v = [x_1 ; x_2 ; \ldots ; x_k] .$$

3.1 Remark. The product of a vector a and a k-vector v is a sum

$$av = a.v + a{\dot{}}v$$

where a.v and $a\dot{}v$ are homogeneous of respective degrees k-1 and k+1. □

3.2 Stokes' theorem. Let f be a continuously differentiable function in a region containing S in its interior. Then

$$\int_S (dS.P)f = \int_{\partial S} ds\, f . \qquad \square$$

The well-known tensor formulation of Stokes' theorem is valid only for the k-vector valued functions. However, the above Clifford algebraic formulation of Stokes' theorem is valid also for the spinor valued functions, and so it concerns all the basic representations of the group $SO(n) \simeq Spin(n)/Z_2$.

3.3 Example. For a vector a we have $(a.P)f(x) = f'(x)a$. Hence, the line integral over a path S

$$\int_S (dx.P)f(x) = \int_S f'(x)dx = \int_{\partial S} f(x)$$

is also formally obtained by Stokes' theorem. It should be noted that this line integral is in general non-zero, even though $f(x)$ would satisfy $Pf(x) = 0$. □

3.4 Example. Let v be a differentiable vector valued function and b an arbitrary bivector. Then $P^{\wedge}v$ is a bivector and $b(P^{\wedge}v)$ is a sum of a 0-vector $b_0(P^{\wedge}v)$, a 2-vector $b_2(P^{\wedge}v)$, and a 4-vector $b_4(P^{\wedge}v)$. Comparing the the homogeneous components of the right hand side and the left hand side of $(bP)v = b(Pv)$ we find that $(b.P).v = b_0(P^{\wedge}v)$. Let now v be irrotational, that is, $P^{\wedge}v = 0$, then $(b.P).v = 0$ for any bivector b. Hence, integrating over a surface S and considering the homogeneous components we obtain by Stokes' theorem

$$\int_S (dS.P).v = \int_{\partial S} dx.v = 0 . \qquad \square$$

From now on $\dim S = n$.

3.5 Example. Let v be a sourceless vector valued function, that is, $P.v = 0$, in a region containing S in its interior. It follows that $(S.P)^{\cdot}v = 0$ and therefore

$$\int_S (dS.P)^{\cdot}v = \int_{\partial S} ds^{\cdot}v = 0 . \qquad \square$$

3.6 Cauchy's theorem. Let f be a regular function in a region containing S in its interior. Then

$$\int_{\partial S} ds\, f = 0 . \qquad \square$$

3.7 Example. Consider a function $q(x-a)$ defined by $q(x) = x|x|^{-n}$. Then

$$\int_{\partial S} ds\; q(x-a) = 0 ,$$

if a lies outside of S. On the contrary, if a lies inside of S, then

$$\int_{\partial S} ds\; q(x-a) = v_n e_J ,$$

where $v_n = n \dfrac{\pi^{n/2}}{(n/2)!}$ and $e_J = e_1 e_2 \ldots e_n$. $\square$

Let now $f(x)$ be a differentiable function. Then

$$P(q(x)f(x)) = (Pq(x))f(x) + q(x)(\tilde{P}f(x)) ,$$

where $\tilde{P}$ is a differential operator satisfying $x\tilde{P} = D - M$. If the function $f(x)$ is antiregular, that is, $(D-M)f(x) = 0$, then

$$P(q(x)f(x)) = 0 .$$

Let the origin 0 be inside a region S_0 and let S contain S_0 in its interior. Then

$$\int_{S \setminus S_0} (dS.P)q(x)f(x) = \int_{\partial S} ds\ q(x)f(x) - \int_{\partial S_0} ds\ q(x)f(x) = 0.$$

If we let S_0 shrink, so that $f(x)$ becomes approximately constant $f(0)$ in S_0, then the last integral tends to $v_n e_J f(0)$. This leads us to the following generalization of a classical result.

3.8 <u>Cauchy's integral formula</u>. Let $f(x)$ be continuous at 0 and antiregular elsewhere in a simply connected domain containing S in its interior. Then

$$\int_{\partial S} ds\ q(x)f(x) = v_n e_J f(0) . \qquad \square$$

In case $n > 2$ a differentiable function $f(x)$ is necessarily constant if it satisfies $(a.P - a^{\bullet}P)f(x) = 0$ for all a in R^n. In case $n = 2$ the condition

$(x.P - x^{\cdot}P)f(x) = 0$ implies $(a.P - a^{\cdot}P)f(x) = 0$ for all a in R^2, and so one may pass from 0 to a in Cauchy's integral formula

$$\int_{\partial S} dx\, q(x-a) f(x) = 2\pi\, e_{12} f(a) .$$

Delanghe et al. [4] have considered (left) regular functions $Pf = 0$ and right regular functions $g\text{P} = 0$, and a generalization of Cauchy's theorem

$$\int_{\partial S} g\, ds\, f = 0$$

and of Cauchy's integral formula

$$\int_{\partial S} u(x-a)\, ds\, f(x) = -(-1)^n s_n e_J f(a)$$

in R^n. Similar results have also been obtained by Hestenes [5].

4. Invariant norm of spinors

In this chapter we shall introduce a linear structure on the space of spinors, define a scalar product of spinors, and show that this scalar product is positive definite. In other words, we shall introduce a norm for spinors. It is important that this norm will be invariant under the rotation group $SO(n)$ of R^n.

For more details on the theory of spinors see, e.g., the works by Chevalley [2] and Crumeyrolle [3].

To fix a minimal left ideal V of R_n we can choose a primitive idempotent f of R_n so that $V = R_n f$. By means of an orthonormal basis $(e_1, e_2, \ldots, e_n)$ for R^n we can construct a primitive idempotent f as follows. Recall that the $\binom{n}{k}$ elements

$$e_A = e_{\alpha_1} e_{\alpha_2} \ldots e_{\alpha_k}, \quad 1 \leq \alpha_1 < \alpha_2 < \ldots < \alpha_k \leq n$$

constitute a basis for the p-vector space R_n^k, and that R_n is the direct sum $R_n = \sum_{k=0}^{n} R_n^k$. Recall also that $\dim_R V = 2^\chi$, where $\chi = [n/2]$ or $\chi = [n/2] + 1$ according as $n = 0,1,2 \bmod 8$ or $n = 3,4,5,6,7 \bmod 8$. There are $n-\chi$ elements e_A, $e_A^2 = 1$, so that they are pairwise commuting and generate a group of order $2^{n-\chi}$, and the idempotent

$$(4.1) \qquad f = \tfrac{1}{2}(1 + e_{A1})\tfrac{1}{2}(1 + e_{A2}) \ldots \tfrac{1}{2}(1 + e_{A(n-\chi)})$$

is primitive, cf. [9].

The division ring $F = f R_n f = \{\psi \in V \mid \psi f = f\psi\}$ is isomorphic to R, C, or H according as $n = 0,1,2 \bmod 8$, $n = 3 \bmod 4$, or $n = 4,5,6 \bmod 8$. The map

$$(4.2) \qquad V \times F \to V, \ (\psi, \lambda) \to \psi\lambda$$

defines a right F-linear structure on V.

The representation $R_n \to \mathrm{End}_F V$, $u \to \tau(u)$ of R_n, where $\tau(u)$ is defined by $\tau(u)\psi = u\psi$, is irreducible. It is also faithful for a simple Clifford algebra, that is, in case $n \neq 1 \bmod 4$.

By the universal property of the Clifford algebra R_n the identity map $x \to +x$ of $R_n^1 = R^n$ is uniquely extended to an anti-involution $u \to u^+$ of R_n. For a k-vector $u \in R_n^k$ this anti-involution is explicitly

$$u^+ = u \quad \text{if} \quad k = 0,1 \bmod 4; \; u^+ = -u \quad \text{if} \quad k = 2,3 \bmod 4.$$

The real linear space

$$P = \{\psi \in V \mid \psi^+ \in V\}$$

has dimension 1, 2, or 4 no matter how large is the dimension of V, cf. [9]. For all $\psi, \varphi \in V$ we have $\psi^+\varphi \in P$. Since moreover $P = F$, the map

$$(4.3) \quad V \times V \to F, \; (\psi,\varphi) \to \psi^+\varphi$$

defines a scalar product on the right F-linear space V. This scalar product is symmetric and non-degenerate, that is, it is positive definite.

The right F-module homomorphisms $V \to V$, $\psi \to s\psi$ which preserve this scalar product, that is, for which $(s\psi)^+(s\varphi) = \psi^+\varphi$ for all ψ,φ in V, form the automorphism group

Aut(V). This automorphism group Aut(V) can be realized as a subgroup of the group

$$G = \{s \in R_n \mid s^+ s = 1\}$$

as follows (according to the simplicity of R_n)

$$G \simeq \mathrm{Aut}(V) \quad \text{when} \quad n \neq 1 \bmod 4$$

$$G \simeq {}^2\mathrm{Aut}(V) \quad \text{when} \quad n = 1 \bmod 4.$$

For $n < 8$ the group G is given by the table

n	G
0	O(1)
1	^{2}O(1)
2	O(2)
3	U(2)
4	Sp(4)
5	^{2}Sp(4)
6	Sp(8)
7	U(8)

and this table continues with a periodicity of 8, cf. [12].

Clearly, if $s \in \mathrm{Spin}(n)$, then $(s\psi)^+(s\varphi) = \psi^+\varphi$ for all ψ,φ in V, and so this scalar product is invariant under the rotation group SO(n) of R^n. Hence, we have an invariant norm of spinors.

If we consider regular or antiregular spinor valued functions, then both the domain R^n and the target $V \subset R_n$ have norms invariant under the rotation group $SO(n)$, and these functions satisfy the corresponding Cauchy's integral formula. Therefore, all those theorems of the classical complex analysis which are based on Cauchy's integral formula and the existence of a norm, e.g., Liouville's theorem and maximum modulus principle, have natural generalizations in higher dimensions.

References

1. R. Bott, The stable homotopy of the classical groups. Ann. of Math. 70 (1959), 313-337.

2. C. Chevalley, The Algebraic Theory of Spinors. Columbia University Press, New York, 1954.

3. A. Crumeyrolle, Algèbres de Clifford et Spineurs. Université Paul Sabatier, Toulouse, 1974.

4. R. Delanghe, F. Brackx, Hypercomplex function theory and Hilbert modules with reproducing kernel. Proc. London Math. Soc. (3) 37 (1978), 545-576.

5. D. Hestenes, Multivector calculus. J. Math. Anal. Appl. 24 (1968), 313-325.

6. M. Karoubi, K-theory. Springer-Verlag, New York, 1978.

7. P. Kustaanheimo, Spinor regularization of Kepler motion. Ann. Univ. Turku AI 73, Publ. Astronom. Obs. Helsinki 102, 1964.

8. P. Kustaanheimo, E. Stiefel, Perturbation theory of Kepler motion based on spinor regularization. J. Reine Angew. Math. 218 (1965), 204-219.

9. P. Lounesto, Sur les idéaux à gauche des algèbres de Clifford et les produits scalaires des spineurs. Ann. Inst. H. Poincaré. Sect. A 33 (1980), 53-61.

10. P. Lounesto, E. Latvamaa, Conformal transformations and Clifford algebras. Proc. Amer. Math. Soc. 79 (1980), 533-538.

11. M. Riesz, Clifford Numbers and Spinors. (Chapters I-IV). Lecture Series No 38. Institute for Physical Science and Technology, Maryland, 1958.

12. C.T.C. Wall, Graded algebras, anti-involutions, simple groups and symmetric spaces. Bull. Amer. Math. Soc. 74 (1968), 198-202.

Contemporary Mathematics
Volume 11, 1982

APPROXIMATE SOLUTIONS OF AN ELLIPTIC EQUATION ON SELECT DOMAINS

Peter A. McCoy
Mathematics Department
United States Naval Academy
Annapolis, Maryland 21402

INTRODUCTION. In classical analytic function theory concerning C^n, certain functions are expanded in terms of series related to their mean boundary values [6-8], approximated by mini-max polynomials in accordance with the Favard-Achieser-Krein theorem [10,15], and interpolated by uniformly convergent sequences of polynomials [9,23]. The present purpose is to carry some of these ideas to series expansions of symmetric Jacobi (ultraspherical) polynomials [1,22] in several variables. The goal is accomplished by extending the function theory on C^n by integral operator methods developed by S. Bergman [3] and R. P. Gilbert [5,11-13].

Expansions of ultraspherical polynomials in one or two variables arise as initial data for the radially symmetric solutions of numerous partial differential equations [4,5,12] including those of the wave, heat, Euler-Poisson-Darboux, and Beltrami. They also occur in the Poisson and Weierstrass kernels [1] and as boundary values for the axially symmetric potential equation [12]. The results contained herein can apply to these problems using maximum and/or uniqueness theorems.

For the n-variable ultraspherical expansion of Φ, we refer to the formal series

$$\Phi(X) = \sum_{K \varepsilon N} A_K \Phi_K(X) \ , \ X = (x_1,\dots,x_n) \ \varepsilon \ E^n \tag{1}$$

$K = (k_1,\dots,k_n) \ \varepsilon \ N := \prod_{j=1}^{n} N$ generated from the ultraspherical polynomials Φ_K that are defined by the Rodriques formula

$$\prod_{j=1}^{n} (1-x_j^2)\Phi_K(x_1,\dots,x_n) = \prod_{j=1}^{n} c_{k_j,\alpha_j} (\partial x_j)^{k_j} (1-x_j^2)^{k_j+\alpha_j} , \alpha_j \geq -1/2$$

$$c_{k,\alpha} = (-1)^k \Gamma(\alpha+1)/2^k \Gamma(k+\alpha+1) \ , \ k = 0,1,\dots \ .$$

The polynomial Φ_K is the eigenfunction

0271-4132/82/0000-0148/$04.50

$$\Lambda\Phi_K(X) = \lambda_K\Phi_K(X) \ , \ X \in S^n := \prod_{j=1}^{n} [-1,+1] \tag{2}$$

of the Gegenbauer's elliptic partial differential operator

$$\Lambda := \sum_{j=1}^{n} \{(1-x_j^2)\partial x_j x_j - 2(\alpha_j+1)x_j\partial x_j\} \tag{3}$$

corresponding to the eigenvalue

$$\lambda_K = -\sum_{j=1}^{n} k_j(k_j+2\alpha_j+1) \ .$$

This follows from eqn.(3) where we observe

$$\Phi_K(X) = R_{k_1}^{\alpha_1}(x_1) \ \dots \ R_{k_n}^{\alpha_n}(x_n) \ , \tag{4}$$

$$R_k^\alpha(x) = P_k^{(\alpha,\alpha)}(x)/P_k^{(\alpha,\alpha)}(1) \ , \ R_k^\alpha(-x) = (-1)^k R_k^\alpha(x)$$

($P_k^{(\alpha,\alpha)}$ are the ultraspherical polynomials [1,22]) and the j-th term in the eqn. (4) is an eigenfunction of the ordinary differential operator Λ (n=1) on [-1,+1] corresponding to the eigenvalue $\lambda_{k_j} = -k_j(k_j+2\alpha_j+1)$.

Consequently, the expansion of Φ with coefficient $A_K := F_K[\lambda_k+\lambda]^{-1}$ solves the Gegenbauer's equation

$$[\Lambda+\lambda]\Phi(X) = F(X) \ , \ \lambda_K+\lambda\neq 0 \tag{5}$$

on S^n when the F has an absolutely and uniformly convergent ultraspherical expansion with coefficients F_K. The Φ may be overdetermined on the ∂S^n. This does not occur when F and hence Φ meet the symmetry condition on S^n:

<u>Case I: n-even (*)</u>.

$$F(x_1,x_2,x_3,\dots,x_{n-1},x_n) = F(x_1,x_2,x_3,\dots,x_{n-1},x_n)_p$$

$$F(-x_1,x_2,x_3,\dots,x_{n-1},x_n) = F(-x_1,x_2,x_3,\dots,x_{n-1},x_n)_s$$

$$F(-x_1,-x_2,x_3,\dots,x_{n-1},x_n) = F(-x_1,-x_2,x_3,\dots,x_{n-1},x_n)_s$$

$$\vdots \qquad\qquad \vdots$$

$$F(-x_1,-x_2,-x_3,\dots,-x_{n-1},x_n) = F(-x_1,-x_2,-x_3,\dots,-x_{n-1},x_n)_s$$

$$F(-x_1,-x_2,-x_3,\dots,-x_{n-1},-x_n) = F(x_1,x_2,x_3,\dots,x_{n-1},x_n)$$

<u>Case II: n-odd (*)</u>.

$$F(x_1,x_2,x_3,\dots,x_{n-1},x_n) = F(x_1,x_2,x_3,\dots,x_{n-1},x_n)_p$$

$$F(-x_1,x_2,x_3,\dots,x_{n-1},x_n) = F(-x_1,x_2,x_3,\dots,x_{n-1},x_n)_s$$

$$F(-x_1,-x_2,x_3,\dots,x_{n-1},x_n) = F(-x_1,-x_2,x_3,\dots,x_{n-1},x_n)_s$$

$$\vdots \qquad\qquad \vdots$$

$$F(-x_1,-x_2,-x_3,\dots,-x_{n-1},x_n) = F(-x_1,-x_2,-x_3,\dots,-x_{n-1},x_n)_s .$$

(*) The subscript p means any permutation of the indices and the subscript s means any permutation of the signs.

Thus, from eqns. (3,5), ultraspherical expansions in polynomials of equal degree ($k_1=\ldots=k_n=k$) solve the symmetric, non-homogeneous Gegenbauer's equation on S^n. If some of the degrees are distinct, the symmetry conditions can be modified. The F-series expansion in ultraspherical polynomials of unequal degree may be determined from the restriction of F to n of the sides of S^n. For the remaining sides, the values of F are fixed as limits taken normal to the side. In fact, we may determine the F-series from its values on the intersection of the union of n-planes $x_k=\gamma_k$ with S^n.

THE MEAN BOUNDARY VALUE PROBLEM. Consider the problem of recapturing Φ from its' arithmetic averages taken over lattice points distributed on the ∂S^n. We approach this by finding a basis for expanding Φ as a series whose coefficients are its' arithmetic averages. To find the basis for expansion, integral transform methods, analogous to those used by R. P. Gilbert [11] in the study of axially symmetric potentials, are applied to associated problems for analytic functions on C^n.

The transform develops from the homogeneous harmonic polynomials

$$H_{k_j}^{\alpha_j}(\xi_j,\eta_j) = \Psi_{k_j}^{\alpha_j}(r_j,x_j) = r_j^{k_j} R_{k_j}^{\alpha_j}(x_j) , \tag{6}$$

$$\xi_j = r_j x_j , \quad \eta_j = \pm r_j\sqrt{1-x_j^2} , \quad -1\leq x_j\leq 1 ,$$

$1\leq j\leq n$, and the polynomial product

$$H_K^\alpha(\xi,\eta) = \Psi_K^\alpha(R,X) = H_{k_1}^{\alpha_1}(\xi_1,\eta_1)\ldots H_{k_n}^{\alpha_n}(\xi_n,\eta_n)$$

in terms of which Φ expands as

$$\Phi(\xi,\eta) = \Psi(R,X) = \sum_{K\in N} A_K H_K^\alpha(\xi,\eta) \tag{7}$$

where $\Phi(\xi,\eta)|_{R=1} = \Phi(X)$. Because $H_{k_j}^{\alpha_j}(\xi_j,0) = \xi_j^{k_j}$,

we find that

$$H_K^\alpha(\xi,0) = \xi^K := \xi_1^{k_1}\ldots\xi_n^{k_n}$$

and that Φ has a formal associate

$$H(Z,0) = \Phi(Z,0) = \sum_{K\in N} A_K Z^K, \; Z^K := Z_1^{k_1}\ldots Z_n^{k_n} \tag{8}$$

on the polydisk $D := \prod_{j=1}^{n}\{Z:|Z|\leq 1\}\subset C^n$.

Gegenbauer's integral representation [1] of the H_k^{α} is

$$H_{k_j}^{\alpha_j}(\xi_j,\eta_j) = G_{\alpha_j}(\sigma^{k_j}) = \int_{-1}^{+1} \sigma_j^{k_j} d\mu_{\alpha_j}(s) \ , \tag{9}$$

$$\sigma_j = (\xi_j + i\eta_j \cos s) \ , \ d\mu_{\alpha_j}(s) = c_{\alpha_j}(\sin s)^{2\alpha_j} ds$$

$$G_{\alpha_j}(1) = 1 \ , \ 1 \leq j \leq n \ .$$

The requisite n-variable analog follows. Let

$$\sigma^K := \sigma_1^{k_1} \ldots \sigma_n^{k_n}$$

$$H_K^{\alpha}(\xi,\eta) := G_{\alpha}(\sigma) = G_{\alpha_1}(\sigma_1^{k_1}) \ldots G_{\alpha_n}(\sigma_n^{k_n}) \ , \tag{10}$$

$$G_{\alpha}(1,\ldots,1) = G_{\alpha}(1) = 1 \ .$$

Then

$$\Phi(\xi,\eta) = \sum_{K \varepsilon N} A_K H_K^{\alpha}(\xi,\eta) = \sum_{K \varepsilon N} A_K G_{\alpha}(\sigma^K) = \tag{11}$$

$$G_{\alpha}(\sum_{K \varepsilon N} A_K \sigma^K) := G_{\alpha}(H(\sigma))$$

when the associate H is uniformly convergent on the polydisk. A sufficient condition is

$$H \ \varepsilon \ A_*^{1+\varepsilon} := \{ \sum_{K \varepsilon N} A_K Z^K : |A_K| = O(1/\ K^{1+\varepsilon})\}$$

$$[K]^{1+\varepsilon} := (k_1 \ldots k_n)^{1+\varepsilon} \ , \ \varepsilon > 0.$$

Since the bounds $\quad |H_K(\xi,\eta)| \leq \sup_{X \varepsilon S^n} |H_K(X)| = 1$

lead to $\quad |\Phi(x)| \leq \sum_{K \varepsilon N} |A_K| \ |H_K(X)| \leq \sum_{K \varepsilon N} |A_K| \ ,$

and

$$\Phi \ \varepsilon \ A^{1+\varepsilon} := \{ \sum_{K \varepsilon N} A_K H_K(\xi,\eta) : |A_K| = O(1/\ K^{1+\varepsilon})\}$$

By construction, the map $G_{\alpha}: A_*^{1+\varepsilon} \to A^{1+\varepsilon}$ is onto. The uniqueness and inverse operator follow from orthogonality:

$$\int_{\partial S^n} \Phi_K \Phi_J d\nu_{\alpha}(X) = h_K^{\alpha} \delta_{KJ} \tag{12}$$

$$d\nu_\alpha(X) = \prod_{j=1}^{n} c_{\alpha_j}(1-x_j^2)^{\alpha_j} dx_j \ , \ \delta_{KJ} = \delta_{k_1 j_1}\delta_{k_2 j_2}\cdots\delta_{k_n j_n}$$

$$h_K^\alpha = \prod_{j=1}^{n} a^{2\alpha_j+1}\Gamma(k_j+\alpha_j+1)^2/(2k_j+2\alpha_j+1)\Gamma(k_j+2\alpha_j+1) \ .$$

The inverse map G_α^{-1}, defined by

$$\int_{\partial S^n} K(ZR^{-1},X)\Psi(R,X)dX = H(Z) \ , \tag{13}$$

with kernel $$K(ZR^{-1},X) = \sum_{K\in N} [\Psi_K(ZR^{-1},X)/h_k^\alpha]$$

is valid on compacta of the polydisk where

$$\limsup |(ZR^{-1})^K/h_K^\alpha|^{1/(k_1+..+k_n)} \geq 1 \ . \tag{14}$$

With the suitable transform available, the arithmetic means [8] of Φ are defined as

$$\sigma_K(\Phi) = \sigma_{k_1\ldots k_n}(\Phi) := (1/[K] \sum_{J\leq K} \Phi(X_{J,K})) = \tag{15}$$

$$(1/k_1\ldots k_n) \sum_{j_1\leq k_1} \ldots \sum_{j_n\leq k_n} \Phi(x_{j_1,k_1},\ldots,x_{j_n,k_n})$$

$$x_{j,k} = \cos(2\pi j/k) \ , \ \theta_{j,k} = 2\pi j/k \ .$$

Furthermore, Riemann coefficients [6-8] $\rho_K(\Phi) = \rho_{k_1\ldots k_n}(\Phi)$ are

$$\rho_K(\Phi) = \sigma_{\infty\ldots\infty}(\Phi) = \int_{\partial S^n} \Phi(X')dX' \ , \ K = \underset{\sim}{0} \ . \tag{16}$$

$$\rho_{0,\ldots,0,k_j,0,\ldots 0}(\Phi) = \sigma_{\infty\ldots\infty,k_j,\infty\ldots\infty}(\Phi) - \sigma_{\infty\ldots\infty}(\Phi), \ k,j\geq 1$$

$$\rho_{0,..,0,k_j,0,\ldots,k_\ell,0,,0}(\Phi) = \sigma_{\infty,..,\infty,k_j,\infty,\ldots,\infty,k_\ell,\infty,\ldots\infty}(\Phi) -$$

$$\sigma_{\infty,\ldots,\infty,k_j,\infty,..\infty}(\Phi) + \sigma_{\infty,\ldots,\infty}(\Phi) - \sigma_{\infty,..,\infty,k_\ell,\infty,\ldots,\infty}(\Phi)$$

$$k_\ell,k_j\geq 1$$

$$\rho_K(\Phi) = \sigma_{k_j',\ldots,k_n'}(\Phi) - \Sigma_1 + \Sigma_2 - \ldots$$

where

$$k_j' = \begin{Bmatrix} k_j \ , \ k_j \neq 0 \\ \infty \ , \ k_j = 0 \end{Bmatrix} ,$$

Σ_K is the sum $\sigma_{k_1',\ldots,\, k_n'}$ with K-non-infinity k_j' replaced by infinity.

Sufficient expansion basis for Φ derives from that of the associate in C^n. The Mobius function appears as

$$\mu(n) := \begin{Bmatrix} 1 \ , \ n = 1 \\ (-1)^k, \ n = q_1 \cdots q_k \\ 0 \ , \ p^2/n \quad (p>1) \end{Bmatrix}$$

and the basis [8] in C^n is

$$p_k(Z) = p_{k_1,\ldots,k_n}(Z_1,\ldots,Z_n) = 1 \ , \ k_1 = \ = k_n = 0$$

$$p_{k_1,\ldots,k_{j-1},0,k_{j+1},\ldots k_n}(Z) = p_{k_1,\ldots,k_{j-1},k_{j+1},\ldots,k_n}(Z_1,\ldots,Z_{j-1},Z_{j+1},\ldots,Z_n)$$

$$p_{k_1,\ldots,k_j}(Z_1,\ldots,Z_j) = \sum_{s_1|k_1} \cdots \sum_{s_k|k_j} \mu(\frac{k_1}{s_1})\ldots\mu(\frac{k_j}{s_j})Z_1^{s_1}\ldots Z_{k_j}^{s_k}$$

$1 \le k_1,\ldots,k_j$, $1 \le k \le n$. Under G_α, the basis is

$$P_K^\alpha(X) = G_\alpha(p_K(\sigma)) = \sum_{s_1|k_1} \cdots \sum_{s_k|k_j} \mu(\frac{k_1}{s_1})\ldots\mu(\frac{k_j}{s_j})\Phi_{s_j}^\alpha(X) \ , \tag{17}$$

$$\Phi_{S_J}(X) = R_{s_1}^{\alpha}(x_1)\ldots R_{s_{k_j}}^{\alpha_j}(X_{k_j}) \ ,$$

on the polydisk leading to

<u>Theorem 1</u>. Let $\Phi^{n-1}(X) = \sum_{k \in N-1} A_K \Phi_K(X) \in A^{1+\varepsilon}$ $(\varepsilon>0)$ on S^{n-1}, and let the associate $\Phi^{n-1}(Z,0)$ analytically continue to the polydisk D. Then the series

$$\Phi(X) = \sum_{K \in N} \rho_K(\Phi^{n-1}) \ P_K^\alpha(X) \tag{18}$$

represents the unique extension Φ of Φ^{n-1} to S^n. Moreover, if $\sigma_K(\Phi^{n-1}) = 0$, $K \in N$, then $\Phi \equiv 0$.

<u>Proof</u>. Construct the sequence of polynomials

$$h_N(Z) = \sum_{K \leq N} \rho_K(\Phi^{n-1})\, p_K(Z) \ , \ N \in N.$$

Because the associate $\Phi^{n-1}(Z,0)$ admits an analytic continuation $\Phi(Z,0)$ to the polydisk D,

$$h_N(Z) \rightrightarrows \Phi(Z,0) = h(Z) \ , \ Z \in D$$

(uniformly) and the restriction of $\Phi(Z,0)$ to $D \cap |Z_n| = 1$ is $\Phi^{n-1}(Z,0)$ by [8,11]. Let us focus on

$$\Phi(X) - \Phi_N(X) = G_\alpha(h(\sigma) - h_N(\sigma)) \ , \ X \in S^n \ .$$

and the subsequent bound

$$|\Phi(X) - \Phi_N(X)| \leq G_\alpha(1)\,||h-h_N|| \ , \ X \in S^n$$

with $||\cdot|| := \sup_D |\cdot|$. Thus,

$$\sup_{X \in S^n} |\Phi(X) - \Phi_N(X)| \leq ||h-h_N|| \to 0 \tag{19}$$

and $\Phi_N \rightrightarrows \Phi$ on S^n. We establish the first part because Φ is uniquely determined by its restriction to the side $X_n = 1$, viz. Φ^{n-1}. Consequently, the vanishing of the arithmetic means is that $\Phi \equiv 0$.

We remark that these expansions are stable because

$$|\sigma_K(\Phi+\varepsilon) - \sigma_K(\Phi)| < \varepsilon \ , \ \varepsilon > 0 \ .$$

When the Φ has symmetry on S^n, a more detailed description follows. The basis

$$Q_k^\alpha(X) := \sum_{j|k} \mu\left(\frac{k}{j}\right) R_j^\alpha(x_1)\ldots R_j^\alpha(x_n) \ , \ X \in S^n \tag{20}$$

$k = 0,1,2,\ldots$ is necessarily symmetric.

<u>Theorem 2</u>. Let $\phi(x_1) = \sum_{k \in N} a_k R_k^\alpha(x_1) \in A^{1+\varepsilon}$ $(\varepsilon > 0)$ on S^1. Then the series,

$$\Phi(X) = \sum_{k \in N} \rho_k(\phi)\, Q_k^\alpha(X) \ , \tag{21}$$

represents the unique extension of ϕ to S^n. If $\sigma_k(\phi) = 0$, $k \in N$, then $\Phi \equiv 0$ on S^n.

Moreover, given $\varepsilon > 0$

$$\sup_{X \in S} |\Phi(X) - \sum_{k=0}^{m} \rho_k(\phi) Q_k^\alpha(X)| \leq M(\delta,\phi)/m^\delta \ ,$$

$m \geq 1$, $0<\delta<\varepsilon$.

<u>Proof</u>. The first part verifies with the precedent reasoning. Note, the basis

simplifies by induction on n and by symmetry. To confirm the estimate in the error bound, apply [8] to the difference $h-h_m$,

$$|\Phi(X) - \sum_{k=0}^{m} a_k\rho_k(\phi)Q_k^{\alpha}(X)| \leq ||h-h_m|| \leq M(\delta,\phi)/m^{\delta}.$$

The next result deals with an interpolating property of symmetric expansions of order $\alpha=-1/2$. The basis in this reduced case becomes

$$Q_k^{-1/2}(X) = Q_k^{-1/2}(\cos\theta_1,\ldots,\cos\theta_n) = \sum_{s/k} \mu(\frac{k}{s})\cos(s\theta_1)\ldots\cos(s\theta_n) \quad (22)$$

$$x_j = \text{arc}\cos\theta_j \quad ,1\leq j\leq n \ , \ k \geq 0 \ .$$

Using properties of the roots of unity, these basic elements interpolate as follows

$$\sigma_m(Q_k^{-1/2}) = \cos(k\theta_1)\ldots\cos(k\theta_{m-1})\cos(k\theta_{m+1})\ldots\cos(k\theta_n)\delta_{nk} \ , \quad (23)$$

leading to

<u>Theorem 3</u>. Let $\{\beta_k\}_{k=0}^{\infty}$ be a real sequence with limit β such that $\beta_k-\beta\varepsilon Q(1/k^{2+\varepsilon})$, $\varepsilon > 0$. Then the ultraspherical series

$$\Phi(X) = \sum_{k=0}^{\infty} (\beta_k-\beta)Q_k^{-1/2}(X) + \beta \ , \ X \ \varepsilon \ S^n \quad (24)$$

is the even function $\Phi \ \varepsilon \ A^{2+\varepsilon}$ for which

$$\sigma_m(\Phi) = \beta_m\cos(m\theta_1)\ldots\cos(m\theta_{m-1})\cos(m\theta_{m+1})\ldots\cos(m\theta_n) \ .$$

On the edge, $\theta_1 = \theta_2 = .. = \theta_{m-1} = \theta_{m+1} = \ldots = \theta_n = 0$, $\sigma_m(\Phi) = \beta_m$, $m = 0,1,2,\ldots$.

<u>Proof</u>. The interpolation is confirmed by eqn. (23) because

$$\sigma_m(\Phi) = \sum_{k=0}^{\infty} (\beta_k-\beta)\sigma_m(Q_k^{-1/2}) + \beta = \beta_m \ , \ m \geq 0$$

with the k-th term in the sum $O(1/k^{2+\varepsilon})$, the series is absolutely and uniformly convergent. Uniqueness follows by applying the result in [8] to the m-th edge and by appealing to symmetry and the maximum principle [14,20]. Finally, $\Phi \ \varepsilon \ A^{2+\varepsilon}$ by the same reasoning that established uniqueness. Note that other interpolation characterizations for normal and tangential derivatives on the faces of S^n can be found from [7,8] in analogous fashion [for example [17,18]].

<u>BEST POLYNOMIAL APPROXIMATION ON S^n.</u> For symmetric, regular Φ on S^n, with

certain bounded derivatives at an edge, we determine the best approximation within a class of ultraspherical polynomials. This may be considered an extension of the classical theorem pertaining to approximation of analytic functions that was discovered by Favard, Achieser and Krein [10,15] and generalized by Babenko [10].

Regard the symmetric form of Φ,

$$\Phi(\xi,\eta) = \sum_{k=0}^{\infty} A_k H_k\,(\xi,\eta)\ ,$$

$$H_k(\xi,\eta) = H_k^{\alpha}(\xi_1,\eta_1)\ldots H_k^{\alpha}(\xi_n,\eta_n)$$

and the subset of regular Φ on S^n for which

$$A_q := \bigcup_{\ell\leq q}\{\Phi : |(\partial\xi_1)^{\ell}\Phi(\xi_1,0,1,0,..,1,0)|\leq 1, |\xi_1|\leq 1\}\ ,$$

$q \geq 1$. Let the polynomial approximating subsets of A_q be

$$\Pi_m := \{P : P(\xi,\eta) = \sum_{k=0}^{m} a_k H_k(\xi,\eta)\}\ ,\ m \geq 0\ .$$

We find the following

Theorem 4. Let the symmetric, ultraspherical series $\Phi \in A_q$, $1 \leq q$. Then,

$$\inf_{P\in\Pi_m} ||\ \Phi - P\ || \leq \Gamma(m-q)/\Gamma(m)\ ,\ q < m\ .$$

Moreover, the bound

$$\sup_{\Phi\in A_q}\ \inf_{P\in\Pi_m}\ ||\ \Phi - P\ || = \Gamma(m-q)/\Gamma(m)\ ,\ q < m$$

is attained for the extremal polynomial Φ_0,

$$\Phi_0(\xi,\eta) = [\Gamma(m-q)/\Gamma(m)]H_m(\xi_1,\eta_1)\ldots H_m(\xi_n,\eta_n) + \Psi(\xi,\eta)$$

where $\Psi \in \Pi_{q-1}$.

Proof. To begin with, symmetry and the maximum principle [14,20] lead to

$$\max_{S^n} |\Phi(\xi,\eta)| = \max_{S^{n-1}} |\Phi(\xi_1,\eta_1,\ldots,\xi_{n-1},\eta_{n-1},1,0)| = \ldots \tag{25}$$

$$= \max_{[-1,+1]} |\Phi(\xi_1,0,1,\ldots,1,0)|.$$

Because $\Phi(Z,0,1,\ldots,1,0) = H(Z)$, the (symmetric) associate has bounded derivative : $|(\partial\xi_1)^\ell H(\xi_1)| \leq 1$, $\ell \leq q$, on the disk $|\xi_1| \leq 1$. And from [10] , follows the approximation

$$\sup_{H \varepsilon A_q^*} \inf_{p \varepsilon \Pi_m^*} ||H-p|| = \Gamma(m-q)/\Gamma(m) ,$$

where the sets

$$A_q^* := \{H: H=G_\alpha^{-1}(\Phi), \ \Phi \ \varepsilon \ A_q\}, \ 1 \leq q$$

and

$$\Pi_m^* := \{p: p = G_\alpha^{-1}(P), \ P \ \varepsilon \ \Pi_m\}, \ m \geq 0.$$

Application of the operator G_α, gives the bounds

$$|\Phi(x)-P(x)| \leq G_\alpha(1,..,1)||H-p|| = ||H-p|| \ , \ x \ \varepsilon \ S^n$$

and

$$\sup_{P \varepsilon \Pi_m} ||\Phi-P|| \leq \sup_{G_\alpha^{-1} P \varepsilon \Pi_m^*} ||H-p|| = \Gamma(m-q)/\Gamma(m) ,$$

$m>q$.

Having determined an upper bound, we now show it to be the least upper bound. In polar coordinates,

$$H_k(\xi,\eta) = r^n R_k^\alpha(x_1)\ldots R_k^\alpha(x_n) ,$$

so as before by the maximum principle, symmetry, and the additional fact that $\Phi_r(x_1,1,..,1)$ is Abel summable to Φ as $r \to 1$ by [1,2],

$$\lim_{r\to 1} \max_{X \varepsilon S^n} |\Phi(r,X)| = \max_{[-1,+1]} |\Phi(x_1,1,..,1)| . \tag{26}$$

From this relation, we see that the stated bound is attained by Φ_0 on S^n. [For further examples see [19]].

INTERPOLATION AND APPROXIMATION. In this section, we indicate a few applications of classical approximation theory in the study of symmetric ultraspherical expansions on S^n. The source of motivation is in a study of axially symmetric potentials found in [16]. Let us begin by constructing a sequence of interpolating polynomials whose convergence will be investigated in turn. For symmetric problems, refer to a point set $C_m = \{(\xi_j,\eta_j)\}_{j=0}^m$, distributed along a side of S^n. Seeking the interpolating polynomial

$$L_m(\xi,\eta) := \sum_{k=0}^m a_k H_k(\xi,\eta) \tag{27}$$

for which

$$L_m(\xi_j,\eta_j) = \Phi(\xi_j,\eta_j) \ , \ 0 \le j \le m \tag{28}$$

along a face, the coefficients a_k are eliminated from the system to give the determinant

$$\Delta(\xi,\eta,C_m) := \begin{vmatrix} L_m(\xi,\eta) & 1 & H_1(\xi,\eta) & \cdots & H_m(\xi,\eta) \\ \Phi(\xi_0,\eta_0) & 1 & H_1(\xi_0,\eta_0) & & H_m(\xi_0,\eta_0) \\ \vdots & \vdots & \vdots & & \vdots \\ \Phi(\xi_m,\eta_m) & 1 & H_1(\xi_m,\eta_m) & & H_m(\xi_m,\eta_m) \end{vmatrix} = 0$$

$$H_k(\xi_j,\eta_j) = H_k(\xi_j,\eta_j,1,0,1,..,1,0) \ , \ 0 \le j, \ k \le m \ .$$

In terms of the [9],

$$V(C_m) = \det|H_k(\xi_j,\eta_j)|_{0 \le j, \ k \le m} \ ,$$

$$V_k(\xi,\eta;C_m) = V(C_m) \mid_{(\xi_k,\eta_k) = (\xi,\eta)}, \ 0 \le k \le m \ .$$

the unique interpolate is

$$L_m(\xi,\eta) = \sum_{k=0}^{m} \Phi(\xi_k,\eta_k) \ V_k \ (\xi,\eta;C_m)/V(C_m) \tag{29}$$

for a distribution of points C_m : $V(C_m) \neq 0$. That is to say that the points (ξ_k,η_k) are distinct and that none of the equations $H_k(\xi_k,\eta_k) = 0$ are satisfied for any j and all k = 0,...,m. For an example, reflect on the zeros of the Tchebyshev polynomials [21] in homogeneous coordinates. Along an edge : $\eta_1 = \xi_2 - 1 = \cdot \cdot \cdot = \eta_m = \xi_m - 1 = 0$, and the polynomial L_m is simply the LaGrange interpolating polynomial [9,16] of the associate. In summary, regard the following

<u>Theorem 5</u>. Let Φ be a regular, symmetric, ultraspherical expansion on S^n, and let $C_m = \{(\xi_j,\eta_j)\}_{j=0}^{m}$, be a point distribution : $V(C_m) \neq 0$ on a side of S^n. Then the unique ultraspherical interpolate of Φ is the polynomial

$$L_m(\xi,\eta) = \sum_{k=0}^{m} \Phi(\xi_k,\eta_k) \ V_k(\xi,\eta;C_m)/V(C_m) \ . \tag{30}$$

Before the problem of convergence, an interpolation of Hermite type [9,21] is considered.The interpolating conditions are

$$L_m^{(j)} \ (\xi_j,\eta_j) = \Phi^{(j)}(\xi_j,\eta_j), \ 0 \le j \le m$$

where the superscript (j) indicates the j-th derivative with respect to the variable ξ_1. Following the reasoning in [16], to eliminate the coefficients in eqn. [27] the Gegenbauer's integral for $H_k(\xi_1,\eta_1)$ is differentiated p-times,

$$(\partial\xi_1)^p\ H_k(\xi_1,\eta_1) = \{\Gamma(k)/\Gamma(k-p)\}H_{k-p}(\xi_1,\eta_1)\ .$$

Continuing in this direction, the determinant in eqn. (28) satisfies the relation

$$(\partial\xi_1)(\partial\xi_2)^2\ldots(\partial\xi_m)^m\ \Delta(\xi,\eta;C_m) = 0 \tag{31}$$

And the cofactor of L_m in eqn. (30) is

$$W(C_m) = (\partial\xi_1)(\partial\xi_2)^2\ldots(\partial\xi_m)^m V(C_m) =$$

$$\det\ [(\partial\xi_j)^j H_k(\xi_j,\eta_j)] = \det\ [(\Gamma(k)/\Gamma(k-j)]H_{k-j}(\xi_j,\eta_j)]$$

$$= 1!2!\ldots m!\ .$$

Defining

$$W_k(\xi,\eta;C_m) = (\partial\xi_1)\ldots(\partial\xi_m)^m V_k(\xi,\eta;C_m)\ ,$$

we find the ultraspherical extension of [16] in

<u>Theorem 6</u>. Let Φ be a regular, symmetric, ultraspherical expansion on S^n, and let $C_m = \{(\xi_j,\eta_j)\}_{j=0}^m$ be a point distribution: $V(C_m) \neq 0$ on a side of S^n. Then the unique interpolate of

$$L_m^{(j)}(\xi_j,\eta_j) = \Phi^{(j)}(\xi_j,\eta_j)\ ,\ 0 \le j \le m$$

is the ultraspherical polynomial

$$L_m(\xi,\eta) = \Phi(\xi_0,\eta_0) + [1!2!\ldots m!]^{-1} \sum_{j=1}^{m} \Phi^{(j)}(\xi_j,\eta_j)W_j(\xi,\eta;C_m)\ . \tag{32}$$

Concerning the question of distributions that produce uniform convergence, the associate of Φ, $H(Z)$, is analytic on $|Z|\le 1$ for regular symmetric Φ on S^n. And, therefore is approximated uniformly by a polynomial sequence [23], $\{H_m\}_{m=1}^\infty$,

$$|H(Z) - H_m(Z)| \le Kr^m\ ,\ |\ Z\ | \le r\ . \tag{33}$$

Now, transform H_m by G_α, $\Gamma_m := G_\alpha(H_m)$, and interpolate the result by L_m^* for each m. With the maximum principle [14,20] on S^n,

$$|\Phi_r(X) - L_{r,m}(X)| \le |\Phi_r(X) - \Gamma_{r,m}(X)| + \tag{34}$$

$$|L_{r,m}(X) - L_{r,m}^*(X)| \le$$

$$\leq C \; ||\Phi_r - \Gamma_{r,m}|| \leq mA||H-H_m|| \leq mAr^m$$

where

$$C = 1 + \max_{(\xi,\eta)\varepsilon S^n} \{ \sum_{k=0}^{m} V_k(\xi,\eta;C_m)/V(C_m)\} \leq 1 + m \, .$$

The subscript r designates the Abel means [1,2] of L_m. From these bounds we find

Theorem 7. Let Φ be a regular, symmetric, ultraspherical expansion on S^n. Then the Abel means

$$\lim_{m\to\infty} L_{r,m}(\xi,\eta) = \Phi_r(\xi,\eta) \;, \; (\xi,\eta) \; \varepsilon \; S^n$$

uniformly with convergence at a geometric rate in r, $0 < r < \rho < 1$. For each fixed m, the interpolates satisfy

$$L_{r,m}(\xi_j,\eta_j) = \Phi_r(\xi_j,\eta_j) \;, \; r = 1, \; 0 \leq j \leq m$$

where $C_m = \{(\xi_j,\eta_j)\}_{j=0}^{m}$ is a point distribution: $V(C_m)\neq 0$ on a side of S^n.

In closing, we observe that a sufficient condition for uniformly convergent interpolates can be given in terms of uniform convergence of the LaGrange interpolates of the associate by the eqns. (8,34). Classical approximation theory limits the choice of interpolating sets. For example, the Lebesgue constants [21] and their connection with the constant C in eqn. (34) shows there is an optimal point distribution that minimizes C. However, there are continuous functions for which C is unbounded as $m\to\infty$ so that in general the ultraspherical interpolating polynomials L_m will not converge uniformly to Φ on the (compact) S^n.

REFERENCES

1. R. Askey, Orthogonal Polynomials and Special Functions, Regional Conference Series in Applied Math., SIAM, Philadelphia, 1975.

2. H. Bavinck, Jacobi Series and Approximation, Mathematical Centre Tracts, No.39, Mathematisch Centrum, Amsterdam, 1972.

3. S. Bergman, Integral Operators in the Theory of Linear Partial Differential Equations, Ergebnisse der Math. und Grenzgebiete, No.23, Springer-Verlag, New York, 1961.

4. L. R. Bragg and J. W. Dettman, Expansions of Solutions of Certain Hyperbolic and Elliptic Problems in Terms of Jacobi Polynomials, Duke J. Math. 36 (1969), 129-144.

5. D. L. Colton, Solution of Boundary Value Problems by the Method of Integral Operators, Research Notes in Math., Vol. 6, Pitman Publishing, San Francisco, 1976.

6. C. H. Ching and C.k. Chui, Uniqueness Theorems Determined by Function Values at the Roots of Unity, J. Approx. Theory 9 (1973), 267-271.

7. ______________, Analytic Functions Characterized by Their Means on an Arc, Trans. Amer. Math. Soc. 184 (1973), 175-183.

8. ______________, Mean Boundary Value Problems and Riemann Series, J. Approx. Theory 19 (1974), 324-336.

9. P. Davis, Interpolation and Approximation, Blaisdell Publishing Co., New York, 1963.

10. S. D. Fisher, Best Approximation by Polynomials, J. Approx. Theory 21 (1977), 43-59.

11. R. P. Gilbert, Function Theoretic Methods in Partial Differential Equations, Math. in Science and Engineering, Vol. 54, Academic Press, New York, 1969.

12. ______________, Constructive Methods for Elliptic Equations, Lecture Notes in Math., Vol. 365, Springer-Verlag, New York, 1974.

13. R. P. Gilbert and R. Newton, eds., Analytic Methods in Mathematical Physics, Gordon and Breach Science Publ., New York, 1970.

14. G. Hellwig, Partial Differential Equations, Blaisdell Publ. Co., New York, 1964.

15. G. G. Lorentz, Approximation of Functions, Holt, Rinehart and Winston, New York, 1966.

16. M. Marden, Axisymmetric Harmonic Interpolation Polynomials in R^n, Trans. Amer. Math. Soc. 196 (1974), 385-403.

17. P. A. McCoy, Mean Boundary Value Problems for a Class of Elliptic Equations in E^3, Proc. Amer. Math. Soc. 76 (1979), 123-128.

18. ______________, A Mean Boundary Value Problem for a Generalized Axisymmetric Potential on a Doubly Connected Region, J. Math. Analysis & Applications, 76 (1980), 213-222.

19. ______________, Mini-Max Approximations of the Dirichlet Problem for a Class of Second Order Elliptic Partial Differential Equations, Subm. publ.

20. B. Muckenhoupt and E. M. Stein, Classical Expansions and their Relation to Conjugate Harmonic Functions, Trans. Amer. Math. Soc. 118 (1965), 17-91.

21. T. J. Rivlin, The Chebyshev Polynomials, John Wiley and Sons, New York, 1974.

22. G. Szegö, Orthogonal Polynomials, Amer. Math. Soc. Colloquium Publ., Vol. 23, Third ed., Providence, 1967.

23. J. L. Walsh, Interpolation and Approximation by Rational Functions in the Complex Domain, Amer. Math. Soc. Colloquium Publ., Vol. 20, Fifth ed., Providence, 1969.

Contemporary Mathematics
Volume 11, 1982

GRADIENT BOUNDS FOR A CLASS OF SECOND ORDER ELLIPTIC EQUATIONS*

by M.H. Protter

INTRODUCTION. Let $u(x)$ for $x \in \Omega \subset R^n$ be a solution of the semilinear equation

$$\Delta u + f(u) = 0 . \tag{1.1}$$

A technique for employing the maximum principle to obtain bounds for the gradient of solutions of (1.1) was developed by Payne [1, 2] and Payne and Stakgold [3, 4]. An essential feature of the method is the derivation of a second order elliptic differential inequality for a combination of the gradient and a functional of f. Since this elliptic inequality may hold even when u itself does not satisfy the maximum principle, for example, when f is positive, it is possible to get gradient bounds when u is an eigenfunction of (1.1) with homogeneous boundary conditions. Denoting the gradient of u by ∇u, we define the function

$$F(x) = |\nabla u(x)|^2 + 2 \int_0^u f(t)\,dt .$$

Then a computation shows that if u is a C^3 solution of (1.1), the function F satisfies the differential inequality

$$\Delta F + \sum_{k=1}^{n} \frac{4f(u)\frac{\partial u}{\partial x_k} - \frac{\partial F}{\partial x_k}}{2|\nabla u|^2} \frac{\partial F}{\partial x_k} \geq 0 . \tag{1.2}$$

No conditions are imposed on the sign of f or f'. Hence if all the coefficients of $\partial F/\partial x_k$ are bounded in Ω, then F satisfies a maximum principle. We conclude that the gradient at any interior point is dominated by a combination of the maximum of the gradient on the boundary and u itself. In particular, if $u = 0$ on $\partial\Omega$ and $f \geq 0$ for all u, then the maximum of the gradient must occur on $\partial\Omega$. However, in many cases, especially when the maximum of u in Ω occurs at an interior point, the coefficients of $\partial F/\partial x_k$ are unbounded in Ω. Thus the maximum principle in (1.2) fails whenever $|\nabla u| = 0$ in Ω. On the other hand, in such cases

---

*Investigation supported in part by NSF grant No. MCS79-05791.

0271-4132/82/0000-0149/$03.00

the maximum of F at those points where $|\nabla u| \neq 0$ is dominated either by the maximum of F on the boundary or by the value of F at a point where $|\nabla u| = 0$. Hence in all cases bounds for the gradient of u are achieved. The results are especially valuable in those situations where it is known that the maximum of the gradient of u cannot occur on $\partial\Omega$. This possibility is sometimes a direct consequence of the geometry of Ω. For example, if Ω is convex and $u = 0$ on $\partial\Omega$, then it is known that the gradient of u cannot have its maximum on $\partial\Omega$. Thus $F(x)$ is dominated by its value at the interior point of Ω where u achieves its maximum.

An extension of the method just described was given by Payne [2] who showed that the function

$$F_1(x) = g(u)|\nabla u|^2 + 2\int_0^u g(t)f(t)\,dt$$

satisfies a second order elliptic differential inequality for solutions u of (1.1) provided that $g(u)$ is chosen appropriately. In this case the class of domains Ω for which one can conclude that $F_1(x)$ cannot assume its maximum on $\partial\Omega$ is larger than the class for which the corresponding function $F(x)$ must have its maximum at an interior point.

Sperb [5] extended the previous work on the Laplace operator to solutions of second order uniformly elliptic equations of the form

$$(1.3)\quad \sum_{i,j=1}^{n} \frac{\partial}{\partial x_i}\left(a_{ij}(x)\frac{\partial u}{\partial x_j}\right) + \sum_{i=1}^{n} b_i(x)\frac{\partial u}{\partial x_i} + c(x)f(u) = 0$$

where $c(x) > 0$ in Ω. Gradient bounds in this case are obtained by showing that

$$F_2(x) = \frac{1}{c(x)}\sum_{i,j=1}^{n} a_{ij}\frac{\partial u}{\partial x_i}\frac{\partial u}{\partial x_j} + 2\int_0^u f(t)\,dt$$

satisfies a second order elliptic differential inequality in which the coefficients of the terms $\partial F_2/\partial x_i$ may be unbounded. We define

$$|\overline{\nabla} u|^2 = \sum_{i,j=1}^{n} g^{ij}\frac{\partial u}{\partial x_i}\frac{\partial u}{\partial x_j}$$

where g^{ij} is the Riemannian metric induced by the principal part of (1.3). Then whenever $|\overline{\nabla} u| = 0$, the maximum principle for F_2 fails and the analysis is similar to that for the Laplacian. The conditions on the coefficients in (1.3) and on the geometry of the domain required for Sperb's extension of the results of Payne depend heavily on certain differential geometric quantities involving the derivatives of the coefficients in (1.3) and the Riemann tensor (g^{ij}).

In this paper we present a greatly simplified technique for extending the method of Payne to general second order elliptic operators. First we show that a maximum principle of the type described above is valid whenever the derivative of the function f has a positive lower bound for all u. This is the case, for example even in the linear case, $f(u) = cu$ with $c > 0$. Then we give a further extension of the method which relaxes the condition on f'.

2. GRADIENT BOUNDS. We use the usual convention that repeated indices indicate sums from 1 to n and that a subscript, say k, preceded by a comma indicates differentiation with resepct to the kth coordinate function. We consider for $x \in \Omega \subset R^n$ the uniformly elliptic second order differential equation

$$a_{km}(x)u_{,km} + b_k(x)u_{,k} + f(u) = 0 . \tag{2.1}$$

Define the function

$$G(x) = u_{,i}u_{,i} + \delta \int_0^u f(t)dt \tag{2.2}$$

where δ is a positive constant to be determined. We shall suppose that u is a C^3 solution of (2.1). Then a computation yields

$$G_{,k} = 2u_{,i}u_{,ik} + \delta f(u)u_{,k}$$

$$G_{,km} = 2u_{,i}u_{,ikm} + 2u_{,ik}u_{,im} + \delta f(u)u_{,km} + \delta f'(u)u_{,k}u_{,m}$$

Hence

$$a_{km}G_{,km} = 2u_{,i}a_{km}u_{,ikm} + 2a_{km}u_{,ik}u_{,im} + \delta f(u)a_{km}u_{,km} + \delta f'(u)a_{km}u_{,k}u_{,m} . \tag{2.3}$$

Also,

$$G_{,k}G_{,k} = 4u_{,i}u_{,ik}u_{,j}u_{,jk} + 4\delta f(u)u_{,i}u_{,k}u_{,ik} + \delta^2 f^2(u)u_{,k}u_{,k} .$$

We suppose that the coefficients in (2.1) are C^1 so that differentiation of that equation with respect to x_i yields

$$a_{km}u_{,ikm} + a_{km,i}u_{,km} + b_k u_{,ik} + b_{k,i}u_{,k} + f'(u)u_{,i} = 0 . \tag{2.4}$$

We multiply (2.4) by $u_{,i}$ and substitute in the first term on the right of (2.3) getting

$$(2.5)\quad a_{km}G_{,km} = 2a_{km}u_{,ik}u_{,im} - 2a_{km,i}u_{,i}u_{,km} - 2b_k u_{,i}u_{,ik} - 2b_{k,i}u_{,i}u_{,k} - 2f'(u)u_{,i}u_{,i} - \delta f(u)b_k u_{,k} - \delta f^2(u) + \delta f'(u)a_{km}u_{,k}u_{,m}\ .$$

We observe that

$$2\delta f u_{,k}G_{,k} = 4\delta f u_{,i}u_{,k}u_{,ik} + 2\delta^2 f^2 u_{,k}u_{,k}\ .$$

Hence

$$(2.6)\quad \frac{1}{2}\,(2\delta f u_{,k} - G_{,k})G_{,k} = -2u_{,i}u_{,j}u_{,ik}u_{,jk} + \frac{1}{2}\,\delta^2 f^2 u_{,k}u_{,k}\ .$$

Let η_0 and η_1 be the lower and upper ellipticity constants in Ω for the operator (2.1) and let M_0 be a positive constant. From (2.5) and (2.6) we find

$$(2.7)\quad a_{km}G_{,km} + \frac{M_0(2\delta f u_{,k} - G_{,k})}{2|\nabla u|^2}\,G_{,k} \geqslant 2(\eta_0 - M_0)u_{,ik}u_{,ik} + \delta f' a_{km}u_{,k}u_{,m} + \delta f^2(\tfrac{1}{2}\,\delta M_0 - 1) - 2a_{km,i}u_{,i}u_{,km} - 2b_k u_{,i}u_{,ik} - 2b_{k,i}u_{,i}u_{,k} - \delta f b_k u_{,k} - 2f' u_{,i}u_{,i}\ .$$

We assume that the coefficients $a_{km}(x)$, $b_k(x)$ together with all their first derivatives are bounded in $L^\infty(\Omega)$ by a positive constant denoted M_1 . Now we apply Cauchy's inequality as follows:

$$-\delta f b_k u_{,k} \geqslant -\frac{1}{4}\,\delta^2 f^2 M_0 - \frac{M_1^2}{M_0}\,u_{,i}u_{,i}$$

$$(2.8)\quad -2a_{km,i}u_{,i}u_{,km} \geqslant -4\,\frac{M_1^2}{\eta_0}\,u_{,i}u_{,i} - \frac{1}{4}\,\eta_0 u_{,km}u_{,km}$$

$$-2b_k u_{,i}u_{,ik} \geqslant -4\,\frac{M_1^2}{\eta_0}\,u_{,i}u_{,i} - \frac{1}{4}\,\eta_0 u_{,km}u_{,km}\ .$$

Substitution of (2.8) into (2.9) yields

$$(2.9)\quad a_{km}G_{,km} + \frac{M_0(2\delta f u_{,k} - G_{,k})}{2|\nabla u|^2}\,G_{,k} \geqslant (\tfrac{5}{2}\,\eta_0 - 2M_0)u_{,ik}u_{,ik} + [f'(\delta\eta_0 - 2) - \frac{M_1^2}{M_0} - 8\,\frac{M_1^2}{\eta_0} - 2M_1] + \delta f^2(\tfrac{1}{4}\,\delta M_0 - 1)\ .$$

We assume that there is a positive constant c_0 such that $f'(u) \geqslant c_0$ for all u. Then we choose $M_0 = \frac{5}{4}\eta_0$ and select δ so that

$$(2.10) \qquad \delta = \max\left[\frac{16}{5\eta_0}, \frac{1}{\eta_0}\left(2 + \frac{M_1^2}{c_0}\left\{\frac{44}{5\eta_0} + \frac{2}{M_1}\right\}\right)\right].$$

With these choices for M_0 and δ the right side of (2.9) is always nonnegative. We thus have the following result.

THEOREM 1. *Suppose the coefficients of* (2.1) *are* $C^1(\bar{\Omega})$ *and that* u *is a* C^3 *solution of* (2.1) *in* Ω. *If there is a positive number* c_0 *such that* $f'(u) \geqslant c_0$ *for all* u *and if* δ *is given by* (2.10), *then whenever* $|\nabla u| \neq 0$ *in* Ω *the function*

$$G(x) = u_{,i}u_{,i} + \delta\int_0^u f(t)\,dt$$

cannot have a maximum value in Ω *unless* $G(x) \equiv \text{const}$.

The condition on the positivity of f' in Ω is rather restrictive, especially when compared with the absence of any condition on f in the case of the Laplacian. A substantially weaker hypothesis results if instead of $G(x)$, the same analysis is applied to the function

$$(2.11) \qquad H(x) = u_{,i}u_{,i} + \delta\int_0^{\phi(u)} f(t)\,dt$$

where $\phi(u)$ is a function to be determined. A computation yields

$$H_{,k} = 2u_{,i}u_{,ik} + \delta f(\phi)\phi'(u)u_{,k}$$

We set $\bar{f}(u) = f[\phi(u)]$ and obtain

$$H_{,k\ell} = 2u_{,i}u_{,ik\ell} + 2u_{,i\ell}u_{,ik} + \delta\bar{f}\phi'u_{,k\ell} + \delta\bar{f}'\phi'^2 u_{,k}u_{,\ell} + \delta\bar{f}\phi''u_{,k}u_{,\ell}.$$

Hence

$$(2.12) \qquad a_{k\ell}H_{,k\ell} = 2a_{k\ell}u_{,i}u_{,ik\ell} + 2a_{k\ell}u_{,i\ell}u_{,ik} + \delta\bar{f}\phi' a_{k\ell}u_{,k\ell} + (\delta\bar{f}'\phi'^2 + \delta\bar{f}\phi'')a_{k\ell}u_{,k}u_{,\ell}.$$

We differentiate (2.1) with respect to x_i and substitute the result in the right side of (2.12) to get

$$a_{k\ell}H_{,k\ell} = -2a_{k\ell,i}u_{,i}u_{,k\ell} - 2b_k u_{,i}u_{,ik} - 2f' u_{,i}u_{,i} - 2b_{k,i}u_{,i}u_{,k} + 2a_{k\ell}u_{,ik}u_{,i\ell} + \delta\bar{f}\phi'[-b_i u_{,i} - f] + (\delta\bar{f}'\phi'^2 + \delta\bar{f}\phi'')a_{k\ell}u_{,k}u_{,\ell} .$$

We assume that ϕ is chosen so that $\phi' \neq 0$ for all u and that the range of ϕ is such that $\bar{f} \neq 0$ for all values of its argument. Then from the relations

$$\frac{fu_{,k}H_{,k}}{|\nabla u|^2} = \frac{2fu_{,i}u_{,k}u_{,ik}}{|\nabla u|^2} + \delta\bar{f}f\phi'$$

$$-\frac{2fu_{,i}u_{,ik}H_{,k}}{\delta\bar{f}\phi'|\nabla u|^2} = -4\frac{u_{,j}u_{,jk}u_{,i}u_{,ik}f}{\delta\bar{f}\phi'|\nabla u|^2} - 2\frac{fu_{,i}u_{,k}u_{,ik}}{|\nabla u|^2}$$

we find

$$\text{(2.13)} \quad a_{k\ell}H_{,k\ell} + \frac{f\bar{f}\delta\phi' u_{,k} - 2fu_{,i}u_{,ik}}{\delta\bar{f}\phi'|\nabla u|^2} H_{,k}$$

$$\geqslant 2a_{k\ell}u_{,ik}u_{,i\ell} - 2a_{k\ell,i}u_{,i}u_{,k\ell} + \left\{(\delta\bar{f}'\phi'^2 + \delta\bar{f}\phi'')a_{k\ell} - 2f'\delta_{k\ell} - 2b_{k,\ell}\right\}u_{,k}u_{,\ell} - 2b_k u_{,i}u_{,ik} - \delta\bar{f}\phi' b_k u_{,k} - \frac{4f}{\delta\bar{f}\phi}u_{,ik}u_{,ik}$$

where $\delta_{k\ell}$ is the Kronecker symbol and we have used the inequality

$$u_{,i}u_{,j}u_{,ik}u_{,jk} \leqslant |\nabla u|^2 u_{,ik}u_{,ik} .$$

Let ε_1, ε_2, ε_3 be positive constants. Then from Cauchy's inequality we find

$$-2|a_{k\ell,i}u_{,i}u_{,k\ell}| \geqslant -\varepsilon_1 M_1 n^2|\nabla u|^2 - \frac{M_1}{\varepsilon_1}u_{,ik}u_{,ik}$$

$$-2|b_k u_{,i}u_{,ik}| \geqslant -\varepsilon_2 M_1|\nabla u|^2 - \frac{M_1}{\varepsilon_2}u_{,ik}u_{,ik}$$

$$-|b_i u_{,i}| \geqslant -\frac{1}{2}n\varepsilon_3 M_1 - \frac{1}{2\varepsilon_3}M_1|\nabla u|^2$$

Hence (2.13) yields the inequality

$$LH \equiv a_{k\ell}H_{,k\ell} + \frac{\delta f\bar{f}\phi' u_{,k} - 2fu_{,i}u_{,ik} - n\varepsilon_3 M_1 u_{,i}u_{,ik} + \frac{1}{2}n\varepsilon_3 M_1\delta\bar{f}\phi' u_{,k}}{\delta\bar{f}\phi'|\nabla u|^2} H_{,k}$$

$$\geqslant 2a_{k\ell}u_{,ik}u_{,i\ell} - \varepsilon_1 M_1 n^2|\nabla u|^2 - \frac{M_1}{\varepsilon_1}u_{,ik}u_{,ik} + [\delta(\bar{f}'\phi'^2 + \bar{f}\phi'')a_{k\ell} - 2f'\delta_{k\ell} - 2|b_{k,\ell}|]u_{,k}u_{,\ell}$$

$$- \varepsilon_2 M_1 |\nabla u|^2 - \frac{M_1}{\varepsilon_2} u_{,ik} u_{,ik} - \frac{1}{2\varepsilon_3} M_1 \delta |\bar{f}\phi'| |\nabla u|^2$$

$$- \frac{2n\varepsilon_3 M_1}{\delta |\bar{f}\phi'|} u_{,ik} u_{,ik} - \frac{4f}{\delta \bar{f}\phi'} u_{,ik} u_{,ik} .$$

Thus we get

(2.14)

$$LH \geqslant [2\mu_0 - M_1(\frac{1}{\varepsilon_1} + \frac{1}{\varepsilon_2} + \frac{2n\varepsilon_3}{\delta|\bar{f}\phi'|}) - \frac{4f}{\delta\bar{f}\phi'}] u_{,ik} u_{,ik}$$

$$+ [\delta\mu_0(\bar{f}'\phi'^2 + \bar{f}\phi'' - \frac{2f'}{\delta\mu_0}) - 2n^2\varepsilon_1 M_1 - \varepsilon_2 M_1 - \frac{\delta M_1 |\bar{f}\phi'|}{\varepsilon_3}] u_{,k} u_{,k}$$

where we have assumed that $\bar{f}'\phi'^2 + \bar{f}\phi'' \geqslant 0$. We suppose further that ϕ is chosen so that $|\bar{f}\phi'| > 0$ for all u . Then we select $\varepsilon_3 = \sqrt{\delta}/|\bar{f}\phi'|$ and (2.14) becomes

(2.15)

$$LH \geqslant [2\mu_0 - M_1(\frac{1}{\varepsilon_1} + \frac{1}{\varepsilon_2} + \frac{2n}{\sqrt{\delta}}) - \frac{4f}{\delta\bar{f}\phi'}] u_{,ik} u_{,ik}$$

$$+ [\delta\mu_0(\bar{f}'\phi'^2 + \bar{f}\phi'' - \frac{2f'}{\delta\mu_0}) - 2n^2\varepsilon_1 M_1 - \varepsilon_2 M_1 - \sqrt{\delta} M_1] u_{,k} u_{,k}$$

The preceding analysis establishes the following result.

THEOREM 2. *Suppose the coefficients of* (2.1) *are* $C^1(\Omega)$ *and that* u *is a* C^3 *solution of* (2.1) *in* Ω . *Suppose there exists a function* ϕ *such that for some positive constant* c_0 *the inequality*

(2.16) $$\bar{f}'\phi'^2 + \bar{f}\phi'' - \frac{2f'}{\delta\mu_0} \geqslant c_0$$

holds for all u *and sufficiently large* δ . *Assume further that* $|\bar{f}\phi'| > 0$ *and that* $|f/\bar{f}\phi'|$ *is bounded for all* u . *Then for sufficiently large* δ , *the function*

$$H(x) = u_{,i} u_{,i} + \delta \int_0^{\phi(u)} f(t)\,dt$$

cannot achieve its maximum in Ω *unless* $H(x) \equiv$ *const.*

PROOF. By choosing ε_1 , ε_2 , δ sufficiently large the first term on the right in (2.15) is nonnegative. Moreover, for fixed ε_1 , ε_2 the quantity

$$\delta\mu_0 c_0 - M_1(2n^2\varepsilon_1 + \varepsilon_2 - \sqrt{\delta})$$

is nonnegative for sufficiently large δ . Hence $LH \geqslant 0$ in Ω

and the result follows.

We observe that for $\phi(u) \equiv u$, the conditions of Theorem 2 reduce to those of Theorem 1. Also if ϕ is chosen so that its range is contained in the portion of the domain of f where f is bounded away from zero, the condition $|f\phi'| > 0$ is easily satisfied. Finally, condition (2.16) holds for sufficiently large ϕ'' and δ provided $\bar{f}'$ is not large compared with $\bar{f}$. In particular if f is a polynomial in u, it is not difficult to see that by selecting ϕ to be a related polynomial, all the hypotheses of Theorem 2 can be satisfied.

BIBLIOGRAPHY

1. Payne, L.E. Bounds for the maximum stress in the Saint Venant torsion problem. Indian J. Mech. Math. (1968) pp. 51-59. Special Issue.

2. ________. Some remarks on the maximum principle, J. Analyse Math. vol. 30(1976) pp. 421-433.

3. Payne, L.E. and I. Stakgold. Nonlinear problems in nuclear reactor analysis, Proc. Conf. on Nonlinear Problems in Phys. Sci and Biol. Lecture Notes (Springer) No. 322 (1972) pp. 298-307.

4. ________________. On the mean value of the fundamental mode in the fixed membrane problem, App. Anal. vol. 3 (1973), pp. 295-303.

5. Sperb, R.P. Maximum principles and nonlinear elliptic problems, J. Analyse Math. vol. 35 (1979), pp. 236-263.

Contemporary Mathematics
Volume 11, 1982

ELLIPTIC SYSTEMS IN THE PLANE ASSOCIATED WITH CERTAIN PARTIAL DIFFERENTIAL EQUATIONS OF DEFORMABLE MEDIA

by

Herbert H. Snyder*

Department of Systems Science and Mathematics
Washington University
Saint Louis, Missouri 63130

and

Department of Mathematics
Southern Illinois University
Carbondale, Illinois 62901

ABSTRACT. The author introduces a family of associative and commutative algebras over the real field, which are associated with linear partial differential equations (in two independent variables x,y) arising in certain problems of the theory of (generally anisotropic) plates and shells. A hypercomplex variable $Z = x + jy$ is introduced, and a class of functions of Z regular in the sense of Riemann. A function f(Z) is shown to be regular if, and only if, its components satisfy a system of generalized Cauchy-Riemann equations. Theorems similar to those of the classical theory are shown to hold when the generalized Cauchy-Riemann system is of elliptic type.

1. Introduction. In several papers (e.g., [1]-[3]) published in the early 1930's, and also in his Hamburg monograph [4] of 1934,

* Current address: RFD A1, Box 7, Cobden, IL 62920, USA

0271-4132/82/0000-0150/$04.25

L. Sobrero introduced hypercomplex function-theoretic methods for the solution of the biharmonic equation,

$$\frac{\partial^4 w}{\partial x^4} + 2\frac{\partial^4 w}{\partial x^2 \partial y^2} + \frac{\partial^4 w}{\partial y^4} = 0 \qquad (1)$$

the fundamental equation in two-dimensional problems of elasticity (of homogeneous and isotropic media). The possibility of extending such methods to higher-order equations with constant coefficients, of the form

$$a_0 \frac{\partial^m w}{\partial x^m} + a_1 \frac{\partial^m w}{\partial x^{m-1} \partial y} + \dots + a_{m-1} \frac{\partial^m w}{\partial x \partial y^{m-1}} + a_m \frac{\partial^m w}{\partial y^m} = 0 \qquad (2)$$

was already noticed in the same period by N. Spampinato [5]. Our own rather extensive search of the technical literature (cf. [14]) shows that there are, indeed, equations of this form of practical importance other than (1). These equations are of orders $m = 4$, 6 and 8, and also appear as the principal equation in various theories of elastic and/or plastic plates and shells. For example, Lundquist and Stowell [6] have considered a fourth-order linear equation with coefficients 1, 2 and 1 in (1) replaced by parameters τ_1, $2\tau_2$ and τ_3, respectively, in which "...the coefficients τ_1, τ_2, and τ_3 allow for the change in the magnitude of the various terms as the plate is stressed beyond the elastic range. In the elastic range $\tau_1 = \tau_2 = \tau_3 = 1$."

For thin plates which are homogeneous but orthotropic, Thielemann [7] found an equation for the deflection $w = w(x,y)$ under load whose associated homogeneous equation may be written

$$D_{11} \frac{\partial^4 w}{\partial x^4} + 4D_{13} \frac{\partial^4 w}{\partial x^3 \partial y} + 2D_{33} \frac{\partial^4 w}{\partial x^2 \partial y^2} + 4D_{23} \frac{\partial^4 w}{\partial x \partial y^3} + D_{22} \frac{\partial^4 w}{\partial y^4} = 0 \qquad (3)$$

where the (constant) coefficients are functions of certain moduli of elasticity and of Poisson's ratio.

Yet another form of anisotropy appears in the theory of the sandwich plate; we refer, in particular, to the study of Libove and Batdorf [8] (who remark inter alia that "the advent of high-speed flight and the concurrent necessity of maintaining aerodynamically smooth surfaces under high stress have led to the increased study of sandwich-plate construction...A sandwich plate consists essentially of a relatively thick, low-density, low-stiffness core bonded between two thin sheets of high-stiffness material.") For small deflections $w = w(x,y)$ of a sandwich plate under load, the authors obtain a complicated linear equation of sixth order. However, the associated homogeneous equation in the principal part alone takes the form

$$A_0 \frac{\partial^6 w}{\partial x^6} + A_1 \frac{\partial^6 w}{\partial x^5 \partial y} + A_2 \frac{\partial^6 w}{\partial x^4 \partial y^2} + \cdots + A_6 \frac{\partial^6 w}{\partial y^6} = 0 \qquad (4)$$

The (constant) coefficients are somewhat intricate functions of three flexural, one twisting, and two shear stiffnesses; two Poisson ratios; and also possibly non-vanishing mid-plane tensile forces. (It is worthy of remark that the last-named forces--in contrast to the elastic case--appear in the coefficients of the principal part.)

Finally, for $m = 8$, Donnell [9] has obtained an equation in the radial deflection of thin cylindrical shells of the form

$$\left(\frac{\partial^2}{\partial x^2} + \frac{\partial^2}{\partial y^2}\right)^4 w + \text{lower order terms} = 0$$

The associated homogeneous equation in the principal part is then the biharmonic operator equation applied twice:

$$\frac{\partial^8 w}{\partial x^8} + 4 \frac{\partial^8 w}{\partial x^6 \partial y^2} + 6 \frac{\partial^8 w}{\partial x^4 \partial y^4} + 4 \frac{\partial^8 w}{\partial x^2 \partial y^6} + \frac{\partial^8 w}{\partial y^8} = 0 \qquad (5)$$

(We have encountered no equation of eighth order containing unsymmetric parital derivatives--terms proportional to $\partial^8 w/\partial x^7 \partial y$, etc.)

Of special interest, then, are homogeneous polynomial operator equations in the plane, of the form (2), in which m takes the values 4 and/or 6, and in which none of the (constant) coefficients vanish, in general. In particular, it will always be assumed that $a_0 \neq 0 \neq a_m$. (In fact, it is convenient to suppose--without loss of generality--that $a_m = 1$; for otherwise, we may divide through by a_m and re-label the coefficients.)

2. Associative Algebras and Regular Functions. For given integral $m \geq 4$, let $\mathfrak{A}_m$ be the associative and commutative algebra over the real field $\mathbb{R}$ spanned by elements

$$1\ (=j^0),\ j,\ j^2,\ \ldots,\ j^{m-1}$$

with $j^p \cdot j^q = j^{p+q}$ for all integral p, q, and such that

$$j^m = -\sum_{k=0}^{m-1} a_k j^k \tag{6}$$

with all real and constant a_k (with $a_0 > 0$). Then an arbitrary element $\alpha \in \mathfrak{A}_m$ takes the form $\alpha = x_0 + x_1 j + x_2 j^2 + \ldots + x_{m-1} j^{m-1}$, $x_k \in \mathbb{R}$. Further, let $P_m(j) = j^m + a_{m-1} j^{m-1} + \ldots + a_1 j + a_0$, the polynomial whose vanishing determines from (6) the multiplication table of $\mathfrak{A}_m$. Define also the function $\|\cdot\| : \mathfrak{A}_m \to \mathbb{R}$ given by $\|\alpha\| = \sum |x_k|$. It is clear that $\|\cdot\|$ defines a norm on $\mathfrak{A}_m$, and computation shows, for any $\alpha, \beta \in \mathfrak{A}_m$, that $\|\alpha\beta\| \leq M\|\alpha\|\,\|\beta\|$ for some constant $M > 0$ (depending on $\mathfrak{A}_m$ but not on α and β).

Let A be the two-dimensional subspace of $\mathfrak{A}_m$ spanned by 1 and j. We shall consider functions $f : A \to \mathfrak{A}_m$, i.e., algebra valued functions f(Z) of the hypercomplex variable $Z = x + jy$. Observe that, as a linear space, A is indistinguishable from $\mathbb{R}^2$. By a domain $\mathcal{D}$ in A, we mean a connected open set in the norm topology. In general, f(Z) is given by an expression of the form

$$f(Z) = \sum_{k=0}^{m-1} j^k f_k(x,y) \tag{7}$$

We shall say that f(Z) <u>is of class</u> $C^{(p)}(\mathcal{D})$--and write $f(Z) \in C^{(p)}(\mathcal{D})$--if the m component functions $f_k(x,y)$ in (7) possess continuous partial derivatives of orders 1, 2, ..., p. Finally, we use the notations: $dZ = dx + j\,dy$; $df = \sum j^k\, df_k$; $\partial f/\partial x = \sum j^k(\partial f_k/\partial x)$, etc., when the expressions in question exist.

Henceforth, we suppose $Z \in \mathcal{D} \subseteq A$ and $f(Z) \in C^{(1)}(\mathcal{D})$. Of the various ways of choosing a class of regular $\mathfrak{A}_m$-valued functions on A from among all $C^{(1)}$-functions, we choose functions which are <u>uniquely differentiable</u> (i.e., regular in the sense of Riemann) and we apply the definition of differentiability first given by G. Scheffers [10]. Hence f(Z) is differentiable on $\mathcal{D}$ if there exists an $\mathfrak{A}_m$-valued function f'(Z) such that df = f'(Z)dZ (where the <u>derivative</u> f'(Z) is independent of dZ). If f(Z) is uniquely differentiable on $\mathcal{D}$, we shall say that f is <u>holomorphic</u> on $\mathcal{D}$. We shall state without proof

<u>Theorem 1</u>. (a) f(Z) <u>is holomorphic on</u> $\mathcal{D}$, <u>with derivative</u>

$$f'(Z) = \frac{\partial f}{\partial x} = \sum j^k \frac{\partial f_k}{\partial x}$$

<u>if and only if</u>,

$$\frac{\partial f}{\partial y} = j\,\frac{\partial f}{\partial x} \tag{8}$$

(b) <u>For</u> m = 4, (8) <u>holds if and only if the component functions</u> f_k <u>of</u> f <u>satisfy the linear system</u> (= generalized Cauchy-Riemann equations for $\mathfrak{A}_4$)

$$\begin{aligned}
\frac{\partial f_0}{\partial y} &= \qquad - a_0 \frac{\partial f_3}{\partial x} \\
\frac{\partial f_1}{\partial y} &= \frac{\partial f_0}{\partial x} - a_1 \frac{\partial f_3}{\partial x} \\
\frac{\partial f_2}{\partial y} &= \frac{\partial f_1}{\partial x} - a_2 \frac{\partial f_3}{\partial x} \\
\frac{\partial f_3}{\partial x} &= \frac{\partial f_2}{\partial x} - a_3 \frac{\partial f_3}{\partial x}
\end{aligned} \tag{9}$$

(c) For $m = 6$, (8) holds if and only if the component functions f_k satisfy system (9) with $\partial f_3/\partial x$ everywhere replaced by $\partial f_5/\partial x$, and augmented by the equations

$$\begin{aligned}
\frac{\partial f_4}{\partial y} &= \frac{\partial f_3}{\partial x} - a_4 \frac{\partial f_5}{\partial x} \\
\frac{\partial f_5}{\partial y} &= \frac{\partial f_4}{\partial x} - a_5 \frac{\partial f_5}{\partial x}
\end{aligned} \tag{10}$$

Our next theorem relates holomorphy to the algebra in a deeper sense.

Theorem 2. Let $f(Z)$ be holomorphic on a domain $\mathcal{D} \subseteq A$, and let $m \geq 4$ be even. Then the associated system of generalized Cauchy-Riemann equations is of elliptic type if and only if $Z = x + jy$ is not a divisor of zero in $\mathfrak{A}_m$.

Proof: (We shall only give here the case $m = 4$, for which all expressions may be given explicitly.) Taking all terms to the left in each equation of (9), we may write (9) as a single matrix-vector equation, in the form

$$A \frac{\partial u}{\partial x} + B \frac{\partial u}{\partial y} = 0 \tag{11}$$

where $u = [f_0 \ \ f_1 \ \ f_2 \ \ f_3]^T$ and

$$A = \begin{bmatrix} 0 & 0 & 0 & a_0 \\ 1 & 0 & 0 & -a_1 \\ 0 & 1 & 0 & -a_2 \\ 0 & 0 & 1 & -a_3 \end{bmatrix}, \quad B = \begin{bmatrix} 1 & 0 & 0 & 0 \\ 0 & -1 & 0 & 0 \\ 0 & 0 & -1 & 0 \\ 0 & 0 & 0 & -1 \end{bmatrix}$$

Then (11) (and hence (9)) is of elliptic type if the polynomial equation in the scalar λ,

$$\det(A - B\lambda) = 0$$

has no real roots (cf. Courant-Hilbert [11]). One finds that $\det(A - B\lambda) = -Q(\lambda)$, where

$$Q(\lambda) = \lambda^4 - a_3\lambda^3 + a_2\lambda^2 - a_1\lambda + a_0 \tag{12}$$

On the other hand, an element $\alpha \in \mathfrak{A}_m$ is a zero-divisor if, and only if, $\det R(\alpha) = 0$, where $R(\alpha)$ is the isomorph of α in the first regular representation of $\mathfrak{A}_m$ by matrices (cf., e.g., Albert [12] or MacDuffee [13]). In particular,

$$R(Z) = \begin{bmatrix} x & 0 & 0 & -a_0y \\ y & x & 0 & -a_1y \\ 0 & y & x & -a_2y \\ 0 & 0 & y & x-a_3y \end{bmatrix}$$

and we obtain

$$\det R(Z) = y^4 \, Q\left(\frac{x}{y}\right) \tag{13}$$

The desired conclusion follows upon comparing (12) and (13), q.e.d. (The theorem is clearly false if m is odd, as one would have expected.)

The polynomial associated with (6) (cf. $P_m(j)$ supra), viz.,

$$P_m(\lambda) = \lambda^m + a_{m-1}\lambda^{m-1} + \ldots + a_2\lambda^2 + a_1\lambda + a_0$$

certainly has no positive zeros, since all $a_k \geq 0$. Moreover, since m is even

$$P_m(-\lambda) = \lambda^m - a_{m-1}\lambda^{m-1} + \ldots + a_2\lambda^2 - a_1\lambda + a_0$$

and $P_m(-\lambda)$ is just proportional to $\det(A - B\lambda)$ in the general case. Hence, if the generalized Cauchy-Riemann system of $\mathfrak{A}_m$ is elliptic, $P_m(\lambda)$ has no negative roots either. It follows that $P_m(j)$ is the product of irreducible quadratic factors. It is the multiplicity of those factors, essentially, which determines the structure of $\mathfrak{A}_m$:

> Theorem 3. Let even $m \geq 4$ be given, and let the generalized Cauchy-Riemann system of $\mathfrak{A}_m$ (e.g., (9) or (10)) be of elliptic type. Then $\mathfrak{A}_m$ is (isomorphic to) either the direct sum of finitely many copies of the complex field, or $\mathfrak{A}_m$ modulo finitely many nilpotent extensions of the complex field is such a direct sum.

For if the factors of $P_m(j)$ are distinct, they are proportional to the orthogonal idempotent elements of $\mathfrak{A}_m$, whose decomposition along which gives the direct sum. If, however, one of the quadratics has multiplicity $p \geq 2$, such a factor corresponds to the complex field supplemented by a nilpotent extension of degree p. The following examples (for m = 4) may prove informative.

Example 1. Let $\mathfrak{A}_4$ be the algebra with

$$P_4(j) = (1 + j^2)(a^2 + j^2) = a^2 + (a^2 + 1)j^2 + j^4 = 0$$

(where the real constant $a^2 \neq 0$ or 1). Then an arbitrary element $Z = x + jy + j^2u + j^3v \in \mathfrak{A}_4$ may be expressed in the form $Z = Z_1e_1 + Z_2e_2$, where the (primitive) orthogonal idempotents e_1 and e_2 are given by

$$e_1 = \frac{1 + j^2}{1 - a^2}, \quad e_2 = -\frac{a^2 + j^2}{1 - a^2}$$

and $Z_1 = x - a^2u + ia(y - a^2v)$, $Z_2 = x - u + i(y - v)$, and where $i = \frac{(a^2 + a + 1)j + j^3}{a(a + 1)}$ satisfies $i^2 = -1$. $\mathfrak{A}_4$ is thus given explicitly as the direct sum of two copies of $\mathbb{C}$.

Example 2. Let $\mathfrak{A}_4$ be the algebra $\mathcal{B}$ of Sobrero (cf. [4]), for which $P_4(j) = (1 + j^2)^2 = 0$, so that the element $\omega = 1 + j^2$ is nilpotent of degree 2. Then $Z = x + jy + j^2u + j^3v \in \mathcal{B}$ has the representation $Z = Z_1 + \omega Z_2$, where $Z_1 = x - u + i(y - v)$, $Z_2 = u + i\,\frac{3v - y}{2}$, and where $i = \frac{3j + j^3}{2}$. Thus $\mathcal{B}$ is a nilpotent extension of $\mathbb{C}$ of degree 2; and $Z = Z_1 + \omega Z_2$ (while almost as useful) is merely a supplementary rather than a direct sum decomposition.

3. Cauchy Integral Theorems. Let $f(Z) \in C^{(1)}(\mathcal{D})$, $\mathcal{D} \subseteq A \subset \mathfrak{A}_m$. If C is a path in $\mathcal{D}$ of points $Z = Z(t) = x(t) + j\,y(t)$, and if $x(t)$ and $y(t)$ are continuous and piecewise-$C^{(1)}$ for all real t in some closed interval, we shall say that C is a smooth path. Forming the product $f(Z)dZ$, we define the line integral of f along C component-wise; in particular for $m = 4$, one obtains

$$\int_C f(Z)dZ = \int_C f_0dx - a_0f_3dy + j\int_C f_1dx + (f_0 - a_1f_3)dy$$
$$+ j^2\int_C f_2dx + (f_1 - a_2f_3)dy + j^3\int_C f_3dx + (f_2 - a_3f_3)dy \tag{14}$$

From (14), the usual basic properties of integrals follow, which we shall not write down here. Instead, we proceed directly to

<u>Theorem 4</u>. <u>Let</u> $f(Z) \in C^{(1)}(\mathcal{D})$, <u>and let</u> $C \subset \mathcal{D}$ <u>be a smooth closed path bounding a region</u> R <u>consisting entirely of points of</u> $\mathcal{D}$. <u>Then</u>

(a) $f(Z)$ <u>is holomorphic on</u> $\mathcal{D}$ <u>if and only if</u>

$$\int_C f(Z)dZ = 0$$

(b) <u>Let</u> $A = a + jb$ <u>be a fixed but arbitrary point of</u> R; <u>then</u>

$$2\pi i\, f(A) = \int_C \frac{f(Z)dZ}{Z - A} + \int\int_R \left(\frac{\partial f}{\partial y} - j\,\frac{\partial f}{\partial x}\right) \frac{dxdy}{Z - A} \tag{15}$$

(c) <u>If</u> $f(Z)$ <u>is holomorphic on</u> $\mathcal{D}$, <u>and given on</u> C, <u>then</u>

$$f(A) = \frac{1}{2\pi i} \int_C \frac{f(Z)dZ}{Z - A} \tag{16}$$

(d) <u>If</u> $G(Z) \in C^{(1)}(\mathcal{D})$, <u>and</u> Z <u>is an arbitrary but fixed point of</u> $\mathcal{D}$ <u>not on</u> C, <u>then</u>

$$g(Z) = \frac{1}{2\pi i} \int_C \frac{G(W)dW}{W - Z} \tag{17}$$

<u>is holomorphic on every domain not meeting</u> C.

Details of proof may be found in [14]. Here, we shall only observe that $\pm\, i$ are those elements of $\mathcal{O}_m$ satisfying $i^2 = -1$; that the double integral in (15) is, in fact, not singular; and lastly, that the left-hand side of (15) is the value of

$$f(A) \int_C \frac{dZ}{Z - A}$$

the evaluation of the integral in which makes use of the structure of $\mathcal{O}_m$ is an essential way.

For our final theorem, we have

<u>Theorem 5</u>. <u>Let</u> $f(Z)$ <u>be holomorphic on a domain</u> $\mathcal{D} \subseteq A \subset \mathfrak{A}_m$. <u>Then every component function</u> $f_k(x,y)$ <u>is a solution on</u> $\mathcal{D}$ <u>of</u>

$$L(w) =$$

$$a_0 \frac{\partial^m w}{\partial x^m} + a_1 \frac{\partial^m w}{\partial x^{m-1}\partial y} + \ldots + a_{m-1} \frac{\partial^m w}{\partial x \partial y^{m-1}} + \frac{\partial^m w}{\partial y^m} = 0$$

<u>Proof</u>: Combining (16) and (17), we conclude that a holomorphic function $f(Z)$ has holomorphic derivatives of all orders. Hence the components have partial derivatives of all orders. Then, by the chain rule,

$$L(f(Z)) = f^{(m)}(Z)\ P_m(j) = 0 = \sum_{k=0}^{m-1} j^k\ L(f_k(x,y)).$$

since L is a linear operator; but then $L(f_k(x,y)) = 0$, q.e.d.

The analysis given to this point is sufficient to show how the general theory of solutions of $L(w) = 0$ is coextensive with the theory of holomorphic functions on the algebras $\mathfrak{A}_m$.

We remark, finally, that our Theorem 4(b) also leads, in a natural way, to the formulation of a theory of generalized holomorphic functions, satisfying equation systems equivalent to

$$\frac{\partial f}{\partial y} - j \frac{\partial f}{\partial x} = \sum_{k=0}^{m-1} j^k \sum_{n=0}^{m-1} - A_{kn}(x,y)\ f_n$$

where $[A_{ij}(x,y)]$ is an $m \times m$ matrix of functions $(i,j = 0,1,\ldots,m-1)$. Substitution from (18) into the double integral in (15) reduces the latter to a system of (non-singular) Fredholm equations for the components of $f(Z)$, which thus continue to satisfy an equation-system of elliptic type.

REFERENCES

1. L. Sobrero: "Di una nova variabile ipercomplessa interessante la teoria dell'elasticita", *Rendiconti della Reale Accademia Nazionale dei Lincei* (6) 19:135-140.

2. ________: "Algebra delle funzioni ipercomplesse e sue applicazioni alla teoria matematica dell'elasticita", *Reale Accademia d'Italia, Memorie della Classe di Scienze fisiche, matematiche e naturali* 6:1-64.

3. ________: "Applicazione degli ipercomplessi ai promlemi di elasticita piana", *Rendiconti della Reale Accademia Nazionale dei Lincei* (6) 19:479-483.

4. ________: *Theorie der Ebenen Elastizität unter Benutzung eines Systems Hyperkomplexer Zahlen.* 17te. Heft, *Hamburger Mathematische Einzelschriften.* B. G. Teubner, Leipzig, 1934.

5. N. Spampinato: "Intorno ad una proprietà delle equazioni differenziali omogenee a coefficienti costanti", *Rendiconti della Reale Accademia Nazionale dei Lincei* (6)21:73-76.

6. E. E. Lundquist and E. Z. Stowell: "Critical compressive stress for flat rectangular plates supported along all edges and elastically restrained against rotation along the unloaded edges", NACA* Report No. 733, USGPO*, Washington, DC, 1942.

7. W. Thielemann: "Contribution to the problem of buckling of orthotropic plates, with special reference to plywood", NACA Tech. Memo. No. 1263, NACA, Washington, DC, 1950.

8. C. Libove and S. B. Batdorf: "A general small-deflection theory for flat sandwich plates", NACA Report No. 899, USGPO, Washington, DC, 1948.

9. L. H. Donnell: "Stability of thin-walled tubes under torsion", NACA Report No. 479, USGPO, Washington, DC, 1933.

10. G. Scheffers: "Verallgemeinerung der Grundlagen der gewohnlich complexen Funktionen", I, II; *Berichte über die Verhandlungen der Sächsischen Akademie der Wissenschaften zu Leipzig, Mathematisch-physikalische Klasse,* 45: 828-848; 46: 120-134.

11. R. Courant and D. Hilbert: *Methods of Mathematical Physics.* vol. II: *Partial Differential Equations.* Interscience Publishers, New York and London, 1962.

12. A. A. Albert: *Structure of Algebras.* American Mathematical Society Colloquium Publications, vol. XXIV. American Mathematical Society, New York, 1939.

13. C. C. MacDuffee: *An Introduction to Abstract Algebra.* John Wiley & Sons, Inc., New York, 1948.

14. H. H. Snyder: *Function-theories on Commutative Algebras Associated with Certain Partial Differential Equations in Mechanics*

of Elastic Media. Indiana Christian University Applied Math. Tech. Rep. No. 81-01, Indianapolis, Indiana. January 1981.

---

*NACA = National Advisory Committee for Aeronautics; USGPO = US Government Printing Office

Contemporary Mathematics
Volume 11, 1982

A Homogeneous Linear PDE in the Plane, With Smooth Real Coefficients, Whose Only Solution is the Zero Function(*)

F. Treves
Department of Mathematics
Rutgers University
New Brunswick, NJ

The present note answers a question of Lee A. Rubel as to whether there exists a real homogeneous linear PDE on some domain of $\mathbb{R}^m$, $P(x,D)U=0$, that has $U=0$ as its only solution. The example given here is that of a fourth-order linear PDE with C^∞ real coefficients in $\mathbb{R}^2$, everywhere elliptic except on the real axis (it is built by means of the Mizohata operator). The solutions under consideration are as general as they can be, namely distributions. The domain of the plane in which they are defined is submitted to the sole demand that it contain the origin.

The argument follows the conceptual line described in Treves [3]. Let $L = \frac{\partial}{\partial y} - iy\frac{\partial}{\partial x}$ denote the Mizohata operator in $\mathbb{R}^2$. We have

$$L\,\bar{L} = \left(\frac{\partial}{\partial y}\right)^2 + y^2\left(\frac{\partial}{\partial x}\right)^2 + i\frac{\partial}{\partial x},$$

$$L\,\bar{L}\,\bar{L}\,L = \left[\left(\frac{\partial}{\partial y}\right)^2 + y^2\left(\frac{\partial}{\partial x}\right)^2\right]^2 + \frac{\partial^2}{\partial x^2},$$

which has real coefficients.

(*) Partly supported by NSF Grant No. 7903545.

We consider a triple sequence $\{K_{m,n,p}\}$ $(m,n,p\in\mathbb{Z}_+)$ of compact subsets $K_{m,n,p}$ of the open upper halfplane $\mathbb{R}^2_+ = \{(x,y);\ y>0\}$, submitted to the following conditions:

(1) the x-projections of the $K_{m,n,p}$ are pairwise disjoint;

(2) for fixed m,n and for $p\to+\infty$, the sets $K_{m,n,p}$ converge to the single point $\{(x_{m,n},y_{m,n})\}$, with $y_{m,n}>0$;

(3) for fixed m, and for $n\to+\infty$, the point $(x_{m,n},y_{m,n})$ converge to (x_m,y_m), with $y_m>0$;

(4) $\lim_{m\to+\infty}(x_m,y_m) = (0,0)$.

We define then a function $\rho\in C_c^\infty(\mathbb{R}^2)$ as follows:

(5) $\rho\geq 0$ everywhere;

(6) $\rho\equiv 0$ in $\mathbb{R}^2\setminus\bigcup_{m,n,p} K_{m,n,p}$;

(7) $\forall\ m,n,p,\ \exists\,\omega_{m,n,p}\in K_{m,n,p}$ such that $\rho(\omega_{m,n,p})>0$.

The idea of using such a function ρ is borrowed from Nirenberg [2].

In the sequel we denote by R an open rectangle $|x|<r$, $|y|<r'$, with the property that, given any m,n,p, either $K_{m,n,p}\subset R$ or $K_{m,n,p}\cap R=\emptyset$. Furthermore we shall call m_R the smallest integer $m_o>0$ such that $(x_{m,n},y_{m,n})\in R$ for infinitely many n, and $(x_m,y_m)\in R$, for all $m>m_o$.

Lemma 1. Let $f,g\in\mathcal{D}'(R)$ be such that f and g are C^∞ functions of $(x,y)\in R$ for $y\neq 0$ and that, in R,

(8) $Lg = \rho f$.

Then, given any $m>m_R$, f vanishes to infinite order at the point (x_m,y_m).

Proof: Set $Z = x + iy^2/2$. Notice that

(9) $LZ = 0.$

For $y>0$ (resp., $y<0$) make the change of variables $X=x$, $Y=y^2/2$. Then

(10) $L = \frac{1}{i}(\frac{\partial}{\partial X} + i\frac{\partial}{\partial Y}).$

It follows from this that, if $\mathcal{O}$ is an open subset either of $\mathcal{R}^+ = \{(x,y)\in\mathcal{R} ; y>0\}$, or of $\mathcal{R}^- = \{(x,y)\in\mathcal{R}; y<0\}$, and if h is any distribution solution of $Lh = 0$ in $\mathcal{O}$, then h is a holomorphic function of Z in $\mathcal{O}$.

This applies to g in (8) if we take either $\mathcal{O}=\mathcal{R}^+\setminus \text{supp}\,\rho$, or $\mathcal{O}=\mathcal{R}^-$. Define then the C^∞ function in $\mathcal{R}^+$, $\check{g}(x,y) = g(x,-y)$. It is a holomorphic function of Z in $\mathcal{R}^+$.

Let now $\mathcal{R}' = \{(x,y)\in\mathcal{R}; |x-x_*|<\delta, |y|<r'\}$ be a subrectangle that does not intersect supp ρ. Any solution (such as g) of $Lh=0$ in $\mathcal{R}'$ is a C^∞ function of y, $|y|<r'$, valued in the space of distributions of x, $|x-x_*|<\delta$, and therefore has a well-defined trace on the segment $|x-x_*|<\delta$, $y=0$. In the case of g this trace must per force be equal to that of $\check{g}$. By uniqueness we conclude that $g=\check{g}$ in $\mathcal{R}'$. But then, since both g and $\check{g}$ are holomorphic functions of Z in $\mathcal{R}^+\setminus\text{supp}\,\rho$ they must be equal in the latter set.

Let now $m>m_{\mathcal{R}}$ and $n,p\in\mathbb{Z}_+$ be large enough that $K_{m,n,p}\subset\mathcal{R}$. We can find a smooth closed curve $\gamma\subset\mathcal{R}$, winding around $K_{m,n,p}$ in $\mathcal{R}^+\setminus\text{supp}\,\rho$, whose inside does not contain any other compact set $K_{m',n',p'}$. We have

(11) $$\oint_\gamma g\,dZ = 0,$$

since g has a holomorphic extension (with respect to Z) to the full inside of γ, namely $\check{g}$. We apply the Stokes formula to (11), taking into account the fact that

(12) $d(gdZ) = -Lg\, dx\wedge dy.$

We get

(13) $$\iint_{K_{m,n,p}} \rho f\, dxdy = 0.$$

Suppose then that $f(x_{m,n},y_{m,n})\neq 0$. As $p\to+\infty$ the arguments of the complex numbers $f(x,y)$, $(x,y)\in K_{m,n,p}$, converge to that of $f(x_{m,n},y_{m,n})$. The argument of $\rho(x,y)\geq 0$ remains equal to zero. And $\rho>0$ in some subset of $K_{m,n,p}$: Eq. (13) is simply not possible. We conclude that $f(x_{m,n},y_{m,n})=0$. Letting then n go to $+\infty$, the conclusion of Lemma 1 follows from (3). □

We consider now a distribution in R, u, such that

(14) $L\,\bar{L}\,\bar{L}\,L\,u - \rho u = 0.$

We know that Eq. (14) has real coefficients. We also know that it is elliptic in R (and more generally in $\mathbb{R}^2$) for $y\neq 0$. Therefore $u\in C^\infty(R^+\cup R^-)$. We have the right to apply Lemma 1 with

$$f = u, \quad g = \bar{L}\,\bar{L}\,L\,u.$$

We conclude that u vanishes to infinite order at every point (x_m,y_m), $m>m_R$.

We observe that the principal symbol of the differential operator $L\,\bar{L}\,\bar{L}\,L$ is the square of that of $L\,\bar{L}$ (see beginning). Consequently we are in a position to apply the strong unique continuation results in Alinhac- Baouendi [1]: since u=0 to infinite order at $(x_m,y_m)\in R^+$ we must have u=0 in R^+.

We use now the fact that u is a C^∞ function of y, $|y|<r'$, valued in the space of distributions of x, $|x|<r$, a property that follows in standard manner from the specific expression of the equation (14) in the coordinates x,y. In particular we see that, if we define $\tilde{u}$ in R by

(15) $\tilde{u}(x,y) = 0$ for $y \geq 0$, $\tilde{u}(x,y) = u(x,y)$ for $y \leq 0$,

we obtain a distribution solution of (14), actually of

(16) $L\,\bar{L}\,\bar{L}\,\tilde{u} = 0$

in R. But uniqueness in the Cauchy problem (across $y=0$) for L, $\bar{L}$ (already used above), implies $\tilde{u}=0$ in R. In summary:

Theorem 1. *The only distribution solution* u *of* Eq. (14) *in* R *is the distribution identically zero.*

Observe that rectangles such a R form a basis of neighborhoods of the origin in $\mathbb{R}^2$.

References

[1] S. Ahsihac and M. S. Baouendi, *Uniqueness for the characteristic Cauchy problem and strong unique continuation for higher order partial differential equations*, Amer. J. Math. 102 (1980), 179-217.

[2] L. Nirenberg, *Lectures on linear partial differential equations*, Reg. Conf. Series in Math., No. 17, Amer. Math. Soc. (1973).

[3] F. Treves, *Remarks about certain first-order linear PDE in two variables*, Comm. P.D.E. 5 (4) (1980), 381-425.

Contemporary Mathematics
Volume 11, 1982

THE NEWTONIAN POTENTIAL FOR A GENERALIZED CAUCHY - RIEMANN OPERATOR IN EUCLIDEAN SPACE

by

F. Brackx and W. Pincket

Seminar of Mathematical Analysis, Seminar of Higher Analysis
State University of Ghent, B-9000 Gent, Belgium

Abstract. We consider the weak and distributional solutions of the non-homogeneous elliptic linear system $D^k u=f$ where D is a generalized Cauchy-Riemann operator in Euclidean space and the functions are Clifford algebra valued. The main tool in this investigation is the so-called Newtonian potential of the operator D^k, which is studied upon its properties with respect to existence, continuity, regularity, integrability, etc.

INTRODUCTION

In a series of papers [1,2,3,4,7,8] R. Delanghe, F. Sommen and one of the authors developed a rather complete function theory for functions defined in Euclidean space of arbitrary dimension and with values in a Clifford algebra. These so-called monogenic functions are the strong solutions of a specific elliptic system of homogeneous linear partial differential equations with constant coefficients, which is related to a generalized Cauchy-Riemann operator D. This theory so provides a direct generalization to higher dimension of the classical theory of holomorphic functions of one complex variable and at the same time a refinenent of harmonic functions.

The aim of this paper is to investigate the weak and distributional solutions of the associated non-homogeneous

0271-4132/82/0000-0152/$07.75

system $D^k u=f$ ($k\in N$) in an open subset Ω of R^{m+1}. This is done by means of a "Newtonian potential" for the operator D^k, which is as well an analogue of the classical Newtonian potential for the Laplacian as another form of the so-called T-operator introduced in the complex case by Vekua in [9] and transposed to higher dimension by Iftimie in [6].

After a short introduction on Clifford-algebras, k-monogenic functions and Clifford-algebra valued distributions, the fundamental solution of the operator D^k is discussed in subsection 2. The third paragraph provides the so-called representation formulae for the weak and distributional solutions of the considered system in an open and connected Ω. These formulae allow to determine the regularity of the solutions accordingly to the right-hand side f. In subsection 4 the Newtonian potential for D^k is introduced and its properties with respect to existence, continuity, regularity, integrability, etc. are established. In a last paragraph it is shown how this Newtonian potential is used to construct solutions of the considered system in an open and bounded Ω.

1. PRELIMINARIES AND NOTATIONS

1.1. Let A be the universal Clifford algebra constructed over a real n-dimensional quadratic vector space V with orthonormal basis $\{e_1,\ldots,e_n\}$. A basis for A is given by

$$\{e_A : A = (h_1,\ldots,h_r)\in P\{1,\ldots,n\} : 1\leq h_1<h_2<\ldots<h_r\leq n\}$$

where $e_\phi = e_0$ is the identity element. Multiplication in A is governed by the following rule for the basic elements :

$$e_i e_j + e_j e_i = -2\delta_{ij}\, e_0 .$$

An involution of A is given by

$$\lambda = \sum_A e_A \lambda_A \rightarrow \bar{\lambda} = \sum_A \bar{e}_A \lambda_A$$

where $\bar{e}_A = (-1)^{\frac{1}{2}n_A(n_A+1)} e_A$, n_A being the cardinality of A.

From the inner product on A

$$(\lambda,\mu)_0 = 2^n \sum_A \lambda_A \mu_A$$

a norm is derived, to wit

$$|\lambda|_0 = 2^{n/2} \left(\sum_A \lambda_A^2\right)^{1/2}$$

which turns A into a Banach algebra.

1.2. If Ω denotes an open subset of Euclidean space R^{m+1} ($m\geq 1$) then functions are considered defined in Ω and taking values in the Clifford algebra A. They are of the form

$$x = (x_0, x_1, \ldots, x_m) \mapsto f(x) = \sum_A e_A f_A(x)$$

where the f_A are, of course, real-valued.

Definition. A function $f\in C_k(\Omega;A)$ is called left-k-monogenic in Ω ($k\in N$) if and only if

$$D^k f = 0 \text{ in } \Omega$$

where the operator D is given by

$$D \doteq \sum_{i=0}^{m} e_i \partial x_i .$$

Notice that in the particular case where $n=m=k=1$ then $A=Re_0 \oplus Re_1 \simeq C$ and $D = e_0 \partial_x + e_1 \partial x_1$ which means that then the holomorphic functions of one complex variable are obtained.

It was shown in [4] that the operator D^k is in fact equivalent to an elliptic system of 2^n homogeneous linear partial differential equations with constant coefficients, which is moreover strongly elliptic in the sense of Garnir [5].

Moreover notice that, as

$$\overline{D}D = D\overline{D} = \Delta_{m+1},$$

Δ_{m+1} being the Laplacian in R^{m+1}, monogenicity implies harmonicity and real-analyticity.

The space of all left-k-monogenic functions in Ω is denoted by $M^{(k)}_{(r)}(\Omega;A)$; it is right A-module.

1.3. Let K be a compact subset of R^{m+1} such that $\overset{\circ}{\bar{K}}=K$. Then the left A-modules

$$\mathcal{D}_{(1)}(K;A) = \{\varphi = \sum_A e_A\varphi_A,\ \varphi_A\in\mathcal{D}(K;R)\}$$

provided with the system of seminorms

$$P_K = \{p_s : s\in N\}$$

where

$$p_s(\varphi) = \sup_{|\beta|\leqslant s}\ \sup_{x\in\mathring{K}} |\partial^\beta\varphi(x)|_0$$

is a Fréchet-Schwartz module.
Now let $K_j\uparrow\Omega$ be an exhausting sequence of compact sets filling up the open set Ω ; then the left A-module of test functions is introduced by

$$\mathcal{D}_{(1)}(\Omega;A) = \lim_{j\in N} \text{ind}\ \mathcal{D}_{(1)}(K_j;A).$$

A left A-distribution in Ω then is a bounded left A-linear functional on $\mathcal{D}_{(1)}(\Omega;A)$, in other words the module of left A-distributions in Ω is the dual module $\mathcal{D}^*_{(r)}(\Omega;A)$ of $\mathcal{D}_{(1)}(\Omega;A)$; it is itself a right A-module :

$$\langle T\lambda+S,\varphi\rangle = \langle T,\varphi\rangle\lambda+\langle S,\varphi\rangle$$

for all $\varphi\in\mathcal{D}_{(1)}(\Omega;A)$ and $\lambda\in A$.
However notice that it is also possible to define $\mu T(\mu\in A)$, again a left A-distribution, by

$$\langle\mu T,\varphi\rangle = \langle T,\varphi\mu\rangle \text{ for all } \varphi\in\mathcal{D}_{(1)}(\Omega;A).$$

Derivatives of A-distributions are defined in the natural way :

$$\langle\partial^\beta T,\varphi\rangle = (-1)^{|\beta|}\langle T,\varphi\rangle \text{ for all } \varphi\in\mathcal{D}_{(1)}(\Omega;A),$$

and so

$$\langle DT,\varphi\rangle = \langle \sum_{i=0}^m e_i\partial_{x_i} T,\varphi\rangle = - \sum_{i=0}^m \langle T,\partial_{x_i}\varphi e_i\rangle$$

$$= -\langle T, \varphi D\rangle$$

and still

$$\langle D^k T, \varphi\rangle = (-1)^k \langle T, \varphi D^k\rangle = \langle T, \varphi(-D)^k\rangle$$

for all $\varphi \in \mathcal{D}_{(1)}(\Omega;A)$.

Finally notice that with $f \in L_1^{loc}(\Omega;A)$ a left A-distribution T_f may be associated by putting

$$\langle T_f, \varphi\rangle = \int \varphi(x)\, f(x) dx^{m+1}, \quad \text{for all } \varphi \in \mathcal{D}_1(\Omega;A).$$

1.4. In the sequel constant use will be made of the following notations.

If $x=(x_o,x_1,\ldots,x_m) \in \mathbb{R}^{m+1}$ then we put

$$x = \sum_{i=o}^{m} e_i x_i,$$

$$\overline{x} = \sum_{i=o}^{m} \overline{e_i}\, x_i = x_o e_o - x_1 e_1 - \ldots - x_m e_m$$

and

$$|x| = |\overline{x}| = \left(\sum_{i=o}^{m} x_i^2\right)^{1/2}.$$

If $\beta=(\beta_o,\ldots,\beta_m)$ is a multi-index, then $|\beta| = \sum_{i=o}^{m} \beta_i$.

Next we have an A-surface element

$$d\sigma = \sum_{i=o}^{m} e_i d\hat{x}_i,$$

where

$$d\hat{x}_i = dx \wedge \ldots \wedge dx_{i-1} \wedge dx_{i+1} \wedge \ldots \wedge dx_m,$$

and a volume element

$$dx^{m+1} = dx_o \wedge \ldots \wedge dx_m.$$

Finally ω_{m+1} will stand for the area of the (m+1)-dimensional unit sphere S_m ; its value is given by

$$\omega_{m+1} = 2\,\pi^{\frac{m+1}{2}} \frac{1}{\Gamma(\frac{m+1}{2})}\ .$$

2. THE FUNDAMENTAL SOLUTION OF THE OPERATOR D^k

2.1. Consider the function E_k given in $R^{m+1}\setminus\{0\}$ by

$$E_k(x) = \frac{1}{\omega_{m+1}}.\frac{1}{(k-1)!}.\frac{\bar{x}x_0^{k-1}}{|x|^{m+1}}.$$

It has the following properties.

(i) $\partial^\beta E_k \in C_\infty(R^{m+1}\setminus\{0\};A)$ for all $|\beta| \leqslant k-1$.

(ii) $\partial^\beta E_k \in L_1^{loc}(R^{m+1};A)$ for all $|\beta| \leqslant k-1$.

(iii) In $R^{m+1}\setminus\{0\}$ holds

$$\partial^j E_k = E_k D^j = E_{k-j} \text{ for } j=0,1,\ldots,k-1$$

while

$$D^k E_k = E_k D^k = 0.$$

(iv) In distributional sense we have

$$D^j E_k = E_k D^j = E_{k-j} \quad \text{for } j=0,1,\ldots,k-1$$

while

$$D^k E_k = E_k D^k = \delta.$$

This distribution E_k is of order zero ; this is closely related to the fact that D^k is a strongly elliptic operator (see also [5]).
In view of the above mentioned properties the function E_k is a fundamental solution of the operator D^k.

2.2. Property (ii) of this fundamental solution E_k is a rather rough statement. Let us have a closer look at the integrability properties of E_k and its derivatives. By p and q always two Hölder conjugated numbers $(\frac{1}{p}+\frac{1}{q}=1)$ are meant.

(A) It is clear that if $k \geqslant m+1$ then

$$E_k \in L_\infty^{loc}(R^{m+1};A).$$

(B) Now assume $k<m+1$ and consider the integral

$$\int_K \frac{|\bar{x}|^q |x_0|^{(k-1)q}}{|x|^{(m+1)q}}\, dx^{m+1}$$

where K is an arbitrary compact set. If $K\subset B(0,R)$ then the considered integral is majorated by

$$\omega_{m+1}\int_0^R r^{m+(k-m-1)q}dr$$

which clearly exists under the condition that

$$(m+1-k)q<m+1$$

or

$$\frac{1}{q}>1-\frac{k}{m+1}$$

or still

$$p>\frac{m+1}{k}>1.$$

This means that if $k<m+1$ then

$$E_k\in L_{q_0}^{loc}(\mathbb{R}^{m+1};A)$$

for all $1\leq q_0<\frac{m+1}{m+1-k}$ or accordingly for all $\frac{m+1}{k}<p_0\leq+\infty$. The same observations can be made for the derivatives of E_k. It is obtained that

$$\partial^\beta E_k\in L_{q_\beta}^{loc}(\mathbb{R}^{m+1};A),\quad 0\leq|\beta|\leq k-1$$

where

$$q_\beta=+\infty \text{ if } k-|\beta|\geq m+1$$

or

$$1\leq q_\beta<\frac{m+1}{m+1-k-|\beta|} \text{ if } k-|\beta|<m+1.$$

3. REPRESENTATION FORMULAE

3.1. Let Ω be an open and connected subset of $\mathbb{R}^{m+1}$. Then we focus our attention to the system

$$D^k u = f \quad \text{in } \Omega \tag{1}$$

where the right-hand side f is at least $L_1^{loc}(\Omega;A)$. Notice that if u is a solution of (1) then it is also a solution of

$$\overline{D}^k D^k u = \overline{D}^k f \text{ in } \Omega$$

or

$$\Delta^k_{m+1} u = f^* \text{ in } \Omega.$$

But as $f^* = \overline{D}^k f$ does not necessarily belong to $L_1^{loc}(\Omega;A)$ anymore, the general theory of elliptic operators with constant coefficients is not relevant to (1) with the above cited restriction on the right-hand side f, making a study of (1) for its own worth-while and so obtaining a refinement for the operator Δ^k_{m+1}.

3.2. Definitions.

(i) The left A-distribution $U \in \mathcal{D}^*_{(r)}(\Omega;A)$ is said to be a *distributional solution* of (1) if for any test function $\varphi \in \mathcal{D}_{(1)}(\Omega;A)$,

$$\langle D^k U, \varphi\rangle = \langle U, \varphi(-D)^k\rangle = \int \varphi(x) f(x) dx^{m+1}.$$

(ii) The function $u \in L_1^{loc}(\Omega;A)$ is said to be a *weak solution* of (1) if for any test function $\varphi \in \mathcal{D}_{(1)}(\Omega;A)$,

$$\int \varphi(-D)^k(x) . u(x) dx^{m+1} = \int \varphi(x) f(x) dx^{m+1}.$$

3.2. Let $\varepsilon>0$; in what follows we make use of the set

$$\Omega_\varepsilon = \{x \in \Omega : d(x, co\Omega) > \varepsilon\}$$

which is open and connected too, and the mollifier $\alpha_\varepsilon \in \mathcal{D}(\mathbb{R}^{m+1};\mathbb{R})$ given by

$$\alpha_\varepsilon(x) = \begin{cases} 1 \text{ if } |x| \leq \varepsilon' < \varepsilon \\ 0 \text{ if } |x| \geq \varepsilon \end{cases}.$$

3.4. The function

$$[(1-\alpha_\varepsilon(y-x))E_k(y-x)]D_y^k$$

clearly belongs to $\mathcal{D}(R^{m+1};A)$ for each y fixed, and has its support in the annular domain

$$\{x : \varepsilon'<|x-y|<\varepsilon\}.$$

If y varies in Ω_ε this support remains contained in Ω, i.e.

$$[(1-\alpha_\varepsilon(y-x))E_k(y-x)]D_y^k\in\mathcal{D}(\Omega;A) \quad \text{for all } y\in\Omega_\varepsilon.$$

So if $U\in\mathcal{D}^*_{(r)}(\Omega;A)$ the expression

$$<U_x,[(1-\alpha_\varepsilon(y-x))E_k(y-x)]D_y^k>$$

is meaningful for all y Ω_ε. Moreover it is infinitely differentiable with respect to y in Ω_ε, differentiation being carried out under the distribution sign.

3.5. The function

$$\alpha_\varepsilon(y-x)E_k(y-x)$$

has, for each y fixed, its support in the spherical domain

$$\{x : |x-y|\leqslant\varepsilon'<\varepsilon\}$$

which remains contained in Ω if y runs through Ω_ε. So, in view of 2.2,

$$\alpha_\varepsilon(y-x)E_k(y-x)\in L_{q_0}^{comp}(\Omega;A) \quad \text{for all } y\in\Omega_\varepsilon,$$

where

$$q_0 = +\infty \quad \text{if} \quad k\geqslant m+1$$

or

$$1\leqslant q_0<\frac{m+k}{m+1-k} \quad \text{if } k<m+1.$$

So the integral

$$\int\alpha_\varepsilon(y-x)E_k(y-x)f(x)dx^{m+1} = \alpha_\varepsilon E_k*f(y)$$

is meaningful *whenever* y *is restricted to* Ω_ε. From the properties of the convolution it then follows that

$$\alpha_\varepsilon E_k * f \in L_{q_0}^{loc}(\Omega_\varepsilon;A)$$

and is even continuous in Ω_ε if $q_0 = +\infty$.
Moreover we have in distributional sense

$$\partial^\beta(\alpha_\varepsilon E_k * f) = \partial^\beta(\alpha_\varepsilon E_k) * f, \quad |\beta| \leqslant k-1$$

where the right-hand side is in $L_{q_\beta}^{loc}(\Omega;A)$.

3.6. <u>Theorem</u>. Let U be a distributional solution of

$$(1) \qquad D^k u = f \text{ in } \Omega, \ f \in L_1^{loc}(\Omega;A) \ ;$$

then there exists a function

$$u \in L_{q_0}^{loc}(\Omega;A)$$

with

$$\partial^\beta u \in L_{q_\beta}^{loc}(\Omega;A), \quad |\beta| \leqslant k-1,$$

such that

$$<U,\varphi> = \int \varphi(x) u(x) dx^{m+1} \quad \text{for all } \varphi \in \mathcal{D}_{(1)}(\Omega;A).$$

This function u is given in Ω_ε, for every $\varepsilon > 0$, by

$$u(y) = <U_x, ((1-\alpha_\varepsilon)E_k(y-x))D_y^k> + \alpha_\varepsilon E_k * f(y), \quad y \in \Omega_\varepsilon$$

where

$$\partial^\beta u(y) = <U_x, \partial^\beta y((1-\alpha_\varepsilon)E_k(y-x))D_y^k> + \partial^\beta(\alpha_\varepsilon E_k) * f(y), \quad y \in \Omega_\varepsilon$$

with $\beta = (\beta_0, \ldots, \beta_m)$ and $|\beta| \leqslant k-1$.

<u>Proof.</u> Take $\varepsilon > 0$ to be fixed and let φ be an arbitrary
test function in $\mathcal{D}_{(1)}(\Omega_\varepsilon;A)$. Then

$$\int_{\Omega_\varepsilon} \varphi(y) u(y) dy^{m+1} = \int \varphi(y) <U_x, ((1-\alpha_\varepsilon)E_k(y-x))D_y^k> dy^{m+1}$$

$$+ \int \varphi(y) (\alpha_\varepsilon E_k * f)(y) dy^{m+1}.$$

Using the fact that E_k is a fundamental solution of D^k, i.e.

$$\int \varphi(y) . E_k(y-x) D_y^k dy^{m+1} = \varphi(x),$$

and the fact that U is a distributional solution of (1), i.e.

$$<U,\chi(-D)^k> = \int\chi(x)f(x)dx^{m+1}$$

with

$$\chi(x) = \varphi*\alpha_\varepsilon E_k(x)$$

which is in $\mathcal{D}_{(1)}(\Omega;A)$ since $\alpha_\varepsilon E_k \in L_{q_0}^{comp}(\mathbb{R}^{m+1};A)$ with $\text{supp}(\alpha_\varepsilon E_k) \subset \{x : |x|<\varepsilon\}$ and $\varphi\in\mathcal{D}_{(1)}(\Omega_\varepsilon,A)$, the first integral on the right-hand side reads successively

$$<U_x,\int\varphi(y).((1-\alpha_\varepsilon)E_k(y-x))D_y^k\ dy^{m+1}>$$

$$= <U_x,\ \int\varphi(y).E_k(y-x)D_y^k\ dy^{m+1}>$$

$$- <U_x,\ \int\varphi(y).(\alpha_\varepsilon E_k(y-x))D_y^k\ dy^{m+1}>$$

$$= <U_x,\varphi(x)>-<U_x,\ (\int\varphi(y)\alpha_\varepsilon(y-x)E_k(y-x)dy^{m+1})(-D_x)^k>$$

$$= <U_x,\varphi(x)> - \int dx^{m+1}(\int\varphi(y)\alpha_\varepsilon(y-x)E_k(y-x)dy^{m+1})f(x)$$

$$= <U_x,\varphi(x)> - \int\varphi(y)dy^{m+1}\int\alpha_\varepsilon(y-x)E_k(y-x)f(x)dx^{m+1}.$$

Hence we obtain

$$\int_{\Omega_\varepsilon}\varphi(y)u(y)dy^{m+1} = <U,\varphi> \text{ for all } \varphi\in\mathcal{D}_{(1)}(\Omega_\varepsilon;A).$$

As $\varepsilon>0$ is arbitrary the above equality applies also to testfunctions in $\mathcal{D}_{(1)}(\Omega;A)$.

The other statements follow immediately from the considerations made in 3.5 where $\varepsilon>0$ is arbitrary.∎

3.7. Corollary. In addition to the above theorem we have the following regularity results.

(i) If $f\in L_\infty^{loc}(\Omega;A)$ then $u\in C_{k-1}(\Omega;A)$;

(ii) if $f\in C_r(\Omega;A)$ then $u\in C_{r+k-1}(\Omega;A)$;

(iii) if $f\in C_\infty(\Omega;A)$ then $u\in C_\infty(\Omega;A)$.

Proof. If $f\in L_\infty^{loc}(\Omega;A)$ then $\partial^\beta(\alpha_\varepsilon E_k)*f$ is continuous in

Ω_ε, for every $\varepsilon>0$, for all $|\beta|\leq k-1$. If $f\in C_r(\Omega;A)$ then $\partial^\beta(\alpha_\varepsilon E_k * f) = \partial^{\beta'}(\alpha_\varepsilon E_k)*\partial^{\beta''} f$ is continuous in Ω_ε, for every $\varepsilon>0$, for all $|\beta'|\leq k-1$ and $|\beta''|\leq r$.∎

3.8. <u>Theorem</u>. If $u\in L_1^{loc}(\Omega;A)$ is a weak solution of (1) then

$$u\in L_{q_0}^{loc}(\Omega;A)$$

with

$$\partial^\beta u\in L_{q_\beta}^{loc}(\Omega;A),\quad |\beta|\leq k-1.$$

Furthermore in Ω_ε, for every $\varepsilon>0$, we have

$$u(y) = \int((1-\alpha_\varepsilon)E_k(y-x))D_y^k.u(x)dx^{m+1}+\alpha_\varepsilon E_k*f(y),\quad y\in\Omega_\varepsilon$$

while analogous formulae hold for the derivatives of u in Ω_ε.

Moreover we have the following regularity results :

(i) if $f\in L_\infty^{loc}(\Omega;A)$ then $u\in C_{k-1}(\Omega;A)$;

(ii) if $f\in C_r(\Omega;A)$ then $u\in C_{r+k-1}(\Omega;A)$ and

(iii) if $f\in C_\infty(\Omega;A)$ then $u\in C_\infty(\Omega;A)$.

<u>Proof</u>. Analogous to the proof of Theorem 3.6.∎

3.9. <u>Remark</u>. The formulae established in Theorems 3.6 and 3.8 link the value of u(y), $y\in\Omega_\varepsilon$ to the "values" of the distribution U or the weak solution u and the function f in the neighbourhood of the point y ; therefore they are called "representation formulae".

4. NEWTONIAN POTENTIALS OF THE OPERATOR D^k

4.1. Starting-point in introducing potentials for the operator D^k is an integral representation of functions of the class C_k in an open subset Ω of R^{m+1}, expressed in the following

<u>Theorem</u>. If $f\in C_k(\Omega;A)$ and S is an (m+1)-dimensional, compact, differentiable oriented manifold-with-boundary

contained in Ω, then

$$\int_{\partial S} \sum_{j=0}^{k-1} (-1)^j E_{j+1}(y-x)d\sigma_y \, D^j f(y)$$

$$+ (-1)^k \int_S E_k(y-x).D^k f(y)dy^{m+1} = \begin{cases} f(x), & x\in\mathring{S} \\ 0 & , x\in coS \end{cases}$$

Proof. First let $x\in coS$. It then follows from Green's Identity [1, Proposition 2.2] that the left-hand side of the stated formula reduces to

$$\int_S E_k(y-x)D_y^k.f(y)dy^{m+1}$$

which apparently equals zero.
Now take $x\in\mathring{S}$ to be fixed and choose $t>0$ such that $\overline{B}(x,t)\subset\mathring{S}$. Then again by Green's Identity we have

$$\int_{\partial(S\setminus B)} \sum_{j=0}^{k-1} (-1)^j E_{j+1}(y-x)d\sigma_y \, D^j f(y)$$

$$= \int_{S\setminus B} (-1)^{k-1} E_k(y-x).D^k f(y)dy^{m+1}.$$

As in the proof of Cauchy's Formula [1, Theorem 2.1] it may be shown that the limit of the left-hand side for $t\to 0+$ is

$$-f(y) + \int_{\partial S} \sum_{j=0}^{k-1} (-1)^j E_{j+1}(y-x)d\sigma_y D_f^j(y).$$

As to the right-hand side observe that $E_k(y-x).D^k f(y)$ is integrable on S since $D^k f$ is continuous on S and

$$\int_S |E_k(y-x)|dy^{m+1} \leq \frac{1}{(k-1)!} \int_0^R r^{k-1}dr,$$

where R is such that $S\subset B(0,R)$. So if $t\to 0+$ then this right-hand side tends to

$$\int_S (-1)^{k-1} E_k(y-x).D^k f(y)dy^{m+1}$$

and the desired result follows.■

4.2. This integral representation enables us to introduce integrals of a special form :

$$\Psi_S^{(j)}(x) = \int_{\partial S} E_{j+1}(x-y)d\sigma_y\, \mu_j(y),\ j=0,1,\dots,k-1$$

and

$$\Phi_S(x) = \int_S E_k(x-y)\rho(y)\, dy^{m+1}.$$

On the analogy of the theory of harmonic functions they are called respectively $(j+1)$-*fold layer potential* $(1\leq j+1\leq k)$ and *Newtonian potential* of the operator D^k with respective *densities* μ_j and ρ.

4.3. Our aim now is to prove some properties of the Newtonian potential

$$\Phi_S = E_k * \rho\delta_S$$

where the density will vary along

$$\mathrm{Lip}_\alpha(S;A)\subset C_0(S;A)\subset L_\infty(S,A)\subset L_p(S,A)\subset L_1(S;A).$$

4.4. Property 1. If $\rho\in L_1(S;A)$ then $\Phi_S\in M_{(r)}^{(k)}(\mathrm{co}S;A)$; if moreover $k<m+1$ then $\lim_{|x|\to+\infty} \Phi_S(x) = 0$.

Proof. Differentiating in distributional sense we have successively

$$\begin{aligned} D^k\Phi_S &= D^k(E_k*\delta_S\rho) = D^kE_k*\delta_S\rho \\ &= \delta*\rho\delta_S \\ &= \rho\delta_S\in L_1(R^{m+1};A). \end{aligned}$$

Now take a testfunction $\varphi\in D_{(1)}(\mathrm{co}S;A)$; then clearly

$$\langle D^k\Phi_S,\varphi\rangle = 0 ;$$

this means that Φ_S is a distributional solution of the equation

$$D^k u = 0 \text{ in } \mathrm{co}S.$$

In view of Corollary 3.7. Φ_S becomes a strong null-solution

of D^k in coS.
If $k<m+1$ then the A-norm of the integrand defining the Newtonian potential is of the form

$$C(n).\frac{1}{|x-y|^m}.|x_0-y_0|^{k-1}|\rho(y)|_0 \leq C(n)\ \frac{|\rho(y)|_0}{|x-y|^\alpha}$$

$$\leq C(n)\ \frac{|\rho(y)|_0}{||x|-|y||^\alpha}$$

where

$$\alpha = m-k+1 > 0.$$

Take $R>0$ sufficiently large in order that $S\subset\mathring{B}(0,R)$; then for all $y\in S$ and $x\in co\overline{B}(0,R)$

$$\frac{|\rho(y)|_0}{||x|-|y||^\alpha} \leq \frac{|\rho(y)|_0}{(|x|-R)^\alpha}.$$

Now give $\varepsilon>0$; choose $|x|>R' = R+\varepsilon^{-1/\alpha}$; then

$$|x|-R > \varepsilon^{-1/\alpha} \text{ or } \frac{1}{(|x|-R)^\alpha} < \varepsilon$$

and so for all $y\in S$ and $|x|>R'(\varepsilon)$ we have

$$|\Phi_S(x)|_0 \leq \frac{1}{\omega_{m+1}}.\frac{1}{(k-1)!}.C(n).\varepsilon\int_S|\rho(y)|_0 dy^{m+1} \leq \varepsilon'.\blacksquare$$

4.5. Remark. An analogous behaviour at infinity holds for the derivatives of Φ_S ; indeed

$$D^j\Phi_S = E_{k-j}*\rho\delta_S, \quad j=0,1,\ldots,k-1$$

and so if $k<m+1+j$ then

$$\lim_{|x|\to+\infty} D^j\Phi_S(x) = 0, \quad j = 0,1,\ldots,k-1.$$

4.6. Before establishing a second property of the Newtonian potential Φ_S we state without proof, which besides is rather classical, a technical lemma.

Lemma. If $f\in L_p(S;A)$ then

$$\lim_{h\to 0} \|f(x+h)-f(x)\|_{L_p(S)} = 0.$$

4.7. Property 2. Assume that $\rho\in L_p(S;A)$.

(i) If $k<m+1$ and $1\leq p\leq\frac{m+1}{k}$ then

$$\Phi_S\in L_{r_0}^{\mathrm{comp}}(R^{m+1};A) \text{ with } \frac{1}{r_0}=\frac{1}{q_0}+\frac{1}{p}-1.$$

(ii) If $k\geq m+1$ and $1\leq p<+\infty$ $\quad(ii_1)$

or $k<m+1$ and $\frac{m+1}{k}<p<+\infty$ $\quad(ii_2)$

then

$$\Phi_S\in C_0(R^{m+1};A)$$

and

$$|\Phi_S(x)|_0\leq M\|\rho\|_{L_p(S)} \text{ for all } x\in S.$$

(iii) If $p=+\infty$ then $\Phi_S\in C_{k-1}(R^{m+1};A)$.

Proof. Under condition (i) the function E_k belongs to $L_{q_0}^{\mathrm{loc}}(R^{m+1};A)$ with $1\leq q_0<\frac{m+1}{m+1-k}$. So, as $\rho\delta_S\in L_p^{\mathrm{comp}}(R^{m+1};A)$, we have

$$\Phi_S=E_k*\rho\delta_S\in L_{r_0}^{\mathrm{loc}}(R^{m+1};A)$$

with $\frac{1}{r_0}=\frac{1}{q_0}+\frac{1}{p}-1$. Notice that $p\leq r<p\frac{m+1}{m+1-kp}$.

Next, under condition (ii_1) the function E_k belongs to $L_\infty^{\mathrm{loc}}(R^{m+1};A)$. So $\Phi_S\in L_\infty^{\mathrm{loc}}(R^{m+1};A)$ and is continuous in R^{m+1}.

As to the case (ii_2) first notice that, in view of Property 1, Φ_S is C_∞ in coS. So we can restrict ourselves to the investigation of the behaviour of Φ_S on S. First we prove that Φ_S is defined all over S. As $k<m+1$ the function E_k belongs to $L_{q_0}^{\mathrm{loc}}(R^{m+1};A)$ with $1\leq q_0<\frac{m+1}{m+1-k}$. So, as $\frac{m+1}{k}<p<+\infty$, it is possible to find q such that $\frac{1}{p}+\frac{1}{q}=1$ and $E_k\in L_q^{\mathrm{loc}}(R^{m+1};A)$. This means that the product

$$E_k(x-y)\rho(y)\in L_1(S;A)$$

for every $x\in S$, in other words $\Phi_S(x)$ is defined for all $x\in S$. The stated estimate then follows at once from Hölders inequality.

Now we prove that Φ_S is continous on S. Take $x \in S$ arbitrarily and let $x' \in S$ be a point in a δ-neighbourhood of x. Then

$$\Phi_S(x)-\Phi_S(x') = \int_{R^{m+1}} (E_k(y-x)-E_k(y-x'))\rho\delta_S(y)dy^{m+1}$$

$$= \int_{B(0,R)} E_k(u)[\rho\delta_S(u+x)-\rho\delta_S(u+x')]du^{m+1}$$

where $R>0$ is chosen in such a way that the bounded set $S-B(x,\delta)$ is contained in $B(0,R)$. So

$$|\Phi_S(x)-\Phi_S(x')|_0$$

$$\leq \frac{C(n)}{\omega_{m+1}(k+1)!} \int_{B(0,R)} |u|^{k-1-m}|\rho\delta_S(u+x)-\rho\delta_S(u+x')|_0 du^{m+1}.$$

Under codition (iii) the function $|u|^{k+1-m}$ belongs to $L_q(B;A)$ with $\frac{1}{p}+\frac{1}{q} = 1$; so Hölder's inequality yields

$$|\Phi_S(x)-\Phi_S(x')|_0 \leq \text{const.}\|\rho\delta_S(u+x)-\rho\delta_S(u+x')\|_{L_p(B)}.$$

In view of Lemma 4.6 we obtain that

$$\lim_{x'\to x} |\Phi_S(x)-\Phi_S(x')|_0 = 0$$

which proves the continuity of Φ_S on S.

For the case (iii) we have in distributional sense that

$$\partial^\beta\Phi_S = \partial^\beta E_k * \rho\delta_S \text{ for all } |\beta|\leq k-1.$$

We always have that $\partial^\beta E_k \in L_1^{loc}(R^{m+1};A)$ for all $|\beta|\leq k-1$. So if $\rho\in L_\infty(S;A)$ then

$$\partial^\beta\Phi_S \in C_0(R^{m+1};A) \text{ for all } |\beta|\leq k-1.$$

This means that $\Phi_S \in C_{k-1}(R^{m+1};A)$.■

4.8. <u>Property 3</u>. If $\rho\in L_p(S;A)$, $1\leq p\leq +\infty$, then the Newtonian potential Φ_S is a weak solution of

$$D^k u = \rho \text{ in } \mathring{S}. \qquad (2)$$

<u>Proof</u>. Taking distributional derivatives we have

$$D^k\Phi_S = D^k(E_k*\rho\delta_S) = D^kE_k*\rho\delta_S$$
$$= \delta*\rho\delta_S = \rho\delta_S$$

which is in $L_p^{comp}(R^{m+1};A)$.

Now take a testfunction $\varphi\in\mathcal{D}(\mathring{S};A)$; then

$$\langle D^k\Phi_S,\varphi\rangle = \langle\rho\delta_S,\varphi\rangle = \langle\rho,\varphi\rangle.$$

This means that Φ_S is a distributional solution of

$$D^k u = \rho \text{ in } \mathring{S}.$$

In view of Theorem 3.6 Φ_S becomes a weak solution of (2). This weak character is $L_{r_0}^{loc}$, C_0 or C_{k-1} accordingly to the conditions (i),(ii) or (iii) of Property 2 respectively.■

4.9. We now state a sufficient condition on the density ρ under which the Newtonian potential Φ_S is a strong solution of the system

$$D^k u = \rho \text{ in } \mathring{S}.$$

<u>Property 4</u>. If $\rho\in Lip_\alpha(S;A)$ then Φ_S is a strong C_k-solution of the system

$$D^k u = \rho \text{ in } \mathring{S}. \qquad (3)$$

<u>Proof</u>. We already know from Properties 3 and 2 (iii) that Φ_S is a weak C_{k-1}-solution of (3). Moreover

$$D^{k-1}\Phi_S = D^{k-1}E_k*\rho\delta_S$$
$$= E_1*\rho\delta_S$$
$$= \frac{1}{\omega_{m+1}}\int_S \frac{\overline{x}-\overline{y}}{|x-y|^{m+1}}.\rho(y)dy^{m+1}$$

which belongs to $C_0(R^{m+1};A)$.

Introduce the function Ψ_S by

$$\Psi_S(x) = \frac{1}{m-1}\frac{1}{\omega_{m+1}}\int_S \frac{1}{|x-y|^{m-1}}\rho(y)dy^{m+1}$$

which is nothing else but the Newtonian potential of the Laplacian in R^{m+1} with density ρ. In view of the assumptions made upon ρ, this $\Psi_S(x)$ satisfies

$$\Delta\Psi_S = \rho \text{ in } \mathring{S},$$

Ψ_S being C_2 in $\mathring{S}$,

But

$$\bar{D}\left(\frac{1}{m-1}\cdot\frac{1}{\omega_{m+1}}\cdot\frac{1}{|x-y|^{m-1}}\right) = \frac{1}{\omega_{m+1}}\cdot\frac{\bar{x}-\bar{y}}{|x-y|^{m+1}}$$

and so, differentiation under the integral sign in the definition of Ψ_S being allowed, we obtain

$$\bar{D}\Psi_S = D^{k-1}\Phi_S$$

which implies that $D^{k-1}\Phi_S$ is still C_1 in $\mathring{S}$, in other words Φ_S is C_k in $\mathring{S}$. Finally we have

$$D^k\Phi_S = D(D^{k-1}\Phi_S) = D(\bar{D}\Psi_S) = \Delta\Psi_S = \rho \text{ in } \mathring{S}. \blacksquare$$

4.10. The Newtonian potential also satisfies certain Hölder continuity conditions. In order to obtain these the following technical lemma is needed.

Lemma. The integral

$$I = \int_S |u-x|^{-\alpha}|u-y|^{-\beta}du^{m+1}$$

where $0\leq\alpha<m+1$, $0\leq\beta<m+1$ and $x,y\in S$, satisfies the following estimates :

$$I \leq M'(\alpha,\beta,m)|x-y|^{m+1-\alpha-\beta} \qquad \text{if } \alpha+\beta>m+1$$

$$I \leq M''(\alpha,\beta,m,S) + C(\alpha,\beta,m)\ |\ln|x-y|| \qquad \text{if } \alpha+\beta=m+1$$

$$I \leq M'''(\alpha,\beta,m,S) \qquad \text{if } \alpha+\beta<m+1.$$

Proof. Analogous to the proof of Lemma 1.1 in [6]. $\blacksquare$

4.11. Property 5. If $\rho\in L_p(S;A)$ with $m+1<p<+\infty$ then Φ_S is Hölder continuous on S with exponent $\alpha=1-\frac{m+1}{p}$.

In the particular case where k=1 the Newtionian potential Φ_S is Hölder continuous in the whole space R^{m+1}.

<u>Proof</u>. We know from Property 2 that under the given assumptions Φ_S is defined and even continuous in the whole space R^{m+1}. So consider the expression

$$F(u,x,y) = \omega_{m+1}(k-1)!\,[E_k(u-x)-E_k(u-y)],\quad u,x,y\in S$$

which can be transformed into

$$F(u,x,y) = |u-x|^{-m}|u-y|^{-m}\,\frac{\overline{u}-\overline{y}}{|u-y|}\cdot$$

$$\left(\frac{|u-y|^{m+1}(u_0-x_0)^{k-1}}{\overline{u}-\overline{y}} - \frac{|u-x|^{m+1}(u_0-y_0)^{k-1}}{\overline{u}-\overline{x}}\right)\frac{\overline{u}-\overline{x}}{|u-x|}.$$

Hence

$$|F(u,x,y)| = |u-x|^{-m}|u-y|^{-m}$$

$$|(u-y)|u-y|^{m-1}(u_0-x_0)^{k-1}-|u-x|^{m-1}(u-x)(u_0-y_0)^{k-1}|$$

while yields

$$|F(u,x,y)| \leq |x-y|\,|u_0-x_0|^{k-1}\sum_{i=0}^{m-1}|u-y|^{i-m}|u-x|^{-1-i}$$

$$+\ |x_0-y_0|\,|u-y|^{-m}\sum_{i=0}^{k-1}|u_0-x_0|^{i}|u_0-y_0|^{k-2-i}$$

and so

$$|\Phi_S(x)-\Phi_S(y)|_0 = |\int_S[E_k(x-u)-E_k(y-u)]\,\rho(u)\,du^{m+1}|_0$$

$$\leq \frac{C(n)\,|x-y|}{\omega_{m+1}(k-1)!}\int_S\left(\sum_{i=0}^{m-1}\frac{|u_0-x_0|^{k-1}}{|u-y|^{m-i}|u-x|^{i+1}}\right.$$

$$\left.+\ \sum_{i=0}^{k-2}\frac{|u_0-x_0|^{i}|u_0-y_0|^{k-2-i}}{|u-y|^{m}}\right)|\rho(u)|_0\,du^{m+1}$$

$$\leq C'|x-y|\left[\sum_{i=1}^{m-1}\int_S\frac{|\rho(u)|_0\,du^{m+1}}{|u-x|^{i+1}|u-y|^{m-i}}+k\int_S\frac{|\rho(u)|_0\,du^{m+1}}{|u-x|\,|u-y|^{m}}\right].\rho(u)$$

Now by Hölder's inequality we get for all $i=0,1,\ldots,m-1$

$$\int_S|\rho(u)|_0\,\frac{1}{|u-x|^{i+1}|u-y|^{m-i}}\,du^{m+1}$$

$$\leq \|\rho\|_{L_p(S)} \left[\int_S \frac{du^{m+1}}{|u-x|^{(i+1)q}|u-y|^{(m-i)q}}\right]^{1/q}.$$

As $m+1<p<+\infty$ we have $1<q<\frac{m+1}{m}$ and so $(i+1)q<m+1$ for all $i=0,1,\ldots,m-1$ while $(i+1)q+(m-i)q=(m+1)q>m+1$. Hence by Lemma 4.10

$$\int_S \frac{du^{m+1}}{|u-x|^{(i+1)q}|u-y|^{(m-i)q}} \leq M'|x-y|^{m+1-(m+1)q}$$

so that

$$\int_S |\rho(u)|_0 \frac{du^{m+1}}{|u-x|^{i+1}|u-y|^{m-i}} \leq M''\|\rho\|_{L_p(S)}|x-y|^{\frac{(m+1)(1-q)}{q}}$$

and

$$|\Phi_S(x)-\Phi_S(y)|_0 \leq C''\|\rho\|_{L_p(S)}|x-y|^{1-\frac{m+1}{p}}$$

where the contant C" depends upon n,m,k,p and S.■

4.12. Property 6. If $\rho\in C_0(S;A)$ then the Newtonian potential Φ_S satisfies the following estimates :

$$|\Phi_S(x)|_0 \leq M \sup_{x\in S}|\rho(x)|_0 \text{ for all } x\in S$$

and

$$|\Phi_S(x)-\Phi_S(y)|_0 \leq M_1 \sup_{x\in S} |\rho(x)|_0 . |x-y| \ln \frac{2R}{|x-y|},$$

for all $x,y\in S$ and where R = diam S.

Proof. From Property 2 we already know that under the stated condition $\Phi_S\in C_0(\mathbb{R}^{m+1};A)$ and

$$|\Phi_S(x)|_0 \leq M'\|\rho\|_{L_p(S)} \quad \text{for all } x\in S,$$

where p is suitably chosen in order to satisfy the conditions (ii_1) or (ii_2). The observation that

$$\|\rho\|_{L_p(S)} \leq (V(S))^{1/p} \sup_{x\in S} |\rho(x)|_0$$

where V(S) stands for the volume of S, leads to the first estimate.

Next we saw in the proof of Property 5 that for all $x,y \in S$,

$$|\Phi_S(x)-\Phi_S(y)|_0 \quad C'|x-y|\left[\sum_{i=1}^{m-1} \int_S \frac{|\rho(u)|_0 du^{m+1}}{|u-x|^{i+1}|u-y|^{m-i}} + k\int_S \frac{|\rho(u)|_0 du^{m+1}}{|u-x||u-y|^m}\right]$$

and so

$$|\Phi_S(x)-\Phi_S(y)|_0 \leq C'|x-y| \sup_{x\in S}|\rho(x)|_0 \left[\sum_{i=1}^{m-1} \int_S \frac{du^{m+1}}{|u-x|^{i+1}|u-y|^{m-i}} + k\int_S \frac{du^{m+1}}{|u-x||u-y|^m}\right].$$

From the proof of Lemma 4.10 we get for all $i=0,1,\dots,m-1$

$$\int_S |u-x|^{-i-1}|u-y|^{-m+i}du^{m+1} \leq 2^m\omega_{m+1} \ln\frac{2R}{|x-y|} + K(m).$$

But $|x-y| \geq R$ and so $\ln\frac{2R}{|x-y|} \geq \ln 2$; hence the above majorations remain valid by substituting $\frac{K(m)}{\ln 2} \ln\frac{2R}{|x-y|}$ for $K(m)$. Finally we obtain

$$|\Phi_S(x)-\Phi_S(y)| \leq C''\sup_{x\in S}|\rho(x)|_0|x-y|\ln\frac{2R}{|x-y|} \text{ for all } x,y\in S,$$

where the contant C" depends upon n,m,k and S.■

4.13. <u>Property 7</u>. If $\rho\in L_\infty(S;A)$ then the Newtonian potential Φ_S satisfies the following estimates :

$$|\Phi_S(x)|_0 \leq M^* \sup_{\substack{x\in S\\ \text{l.a.e.}}} |\rho(x)|_0 \quad \text{for all } x\in S$$

and

$$|\Phi_S(x)-\Phi_S(y)|_0 \leq M_1^* \sup_{\substack{x\in S\\ \text{l.a.e.}}} |\rho(x)|_0 . |x-y|\ln\frac{2R}{|x-y|}$$

for all $x,y\in S$ and where $R = \operatorname{diam} S$.

Proof. Analogous to the proof of Property 6.∎

4.14. Concerning the integrability of the Newtionian potential we can deduce from Property 2 that if $\rho\in L_p(S;A)$ then $\Phi_S\in L_\gamma^{loc}(R^{m+1};A)$ for all $1\leq\gamma\leq+\infty$ whenever $k<m+1$ and $\frac{m+1}{k}<p\leq+\infty$ a $k\geq m+1$ and $1\leq p\leq+\infty$, whereas $\Phi_S\in L_{r_0}^{loc}(R^{m+1};A)$ with $\frac{1}{r_0}=\frac{1}{p}+\frac{1}{q_0}-1$ where $1\leq q_0<\frac{m+1}{m+1-k}$ when $k<m+1$ and $1\leq p\leq\frac{m+1}{k}$.

4.15. In Properties 2,6 and 7 we proved some estimates for Φ_S valid on bounded sets. Now we are looking for such estimates valid all over R^{m+1}. To that end we first prove a technical lemma.

Lemma. Let $\sigma_1>0$, $\sigma_2>0$, $\sigma_1\neq\sigma_2$ and $0\leq\mu\leq1$. Then

$$\frac{|\sigma_1{}^\mu-\sigma_2{}^\mu|}{|\sigma_1-\sigma_2|^\mu}\leq 1.$$

Proof. For $\mu=0$ and $\mu=1$ the assertion is trivial ; so let $\mu\in]0,1[$. Assume that $\sigma_1>\sigma_2$; put $\sigma=\frac{\sigma_2}{\sigma_1}\in[0,1[$. Then

$$\frac{|\sigma_1{}^\mu-\sigma_2{}^\mu|}{|\sigma_1-\sigma_2|}=\frac{1-\sigma^\mu}{(1-\sigma)^\mu}=F(\sigma).$$

As $\frac{dF}{d\sigma}=\mu\frac{(\sigma^{1-\mu}-1)}{(1-\sigma)^{1+\mu}\sigma^{1-\mu}}<0$ and $F(0)=1$ whe have $F(\sigma)\leq1$ for all $\sigma\in[0,1[$.∎

4.16. Property 8. If $\rho\in L_p(S;A)$ with $1<\frac{m+1}{k}<p\leq\frac{m}{k-1}$ then

$$|\Phi_S(x)|_0\leq M\|\rho\|_{L_p(S)}\quad\text{for all } x\in R^{m+1}.$$

If $k=m+1$ the same estimate holds for arbitrary p.

Proof. We know already from Property 2 that under the given assumptions

$$|\Phi_S(x)|_0\leq M\|\rho\|_{L_p(S)}\quad\text{for all } x\in S.$$

So take an arbitrary but fixed $x\in coS$ and put $\delta=dist(x,S)>0$. Then there exists $y_0\in S$ such that $|x-y_0|=\delta$. If $R=diam\ S$

then for any $y \in S$

$$|x-y| \leqslant |x-y_0|+|y_0-y| \leqslant \delta+R$$

and hence

$$S \subset \{z \in \mathbb{R}^{m+1} : \delta \leqslant |x-z| \leqslant \delta+R\} = S'.$$

So

$$\int_S \left(\frac{|\bar{x}-\bar{y}|\,|x_0-y_0|^{k-1}}{|x-y|^{m+1}}\right)^q dy^{m+1} \leqslant \int_{S'} \frac{|\bar{x}-\bar{y}|^q |x_0-y_0|^{(k-1)q}}{|x-y|^{(m+1)q}} dy^{m+1}$$

$$\leqslant \omega_{m+1} \int_{\delta}^{R+\delta} r^{m-(m+1)q+kq} dr$$

$$= \frac{\omega_{m+1}}{m+1-(m+1-k)q}[(R+\delta)^{m+1-(m+1-k)q} - \delta^{m+1-(m+1-k)q}].$$

As $\frac{m+1}{k}<p\leqslant\frac{m}{k-1}$ we have $0<m+1-(m+1-k)q<1$ and so by Lemma 3.13

$$\int_S \left(\frac{|\bar{x}-\bar{y}|\,|x_0-y_0|^{k-1}}{|x-y|^{m+1}}\right)^q dy^{m+1} \leqslant \frac{\omega_{m+1} R^{m+1-(m+1-k)q}}{m+1-(m+1-k)q}.$$

Hence by Hölders inequality

$$|\Phi_S(x)|_0 \leqslant M\|\rho\|_{L_p(S)} \quad \text{for all } x \in \text{co}S$$

with the same constant M, depending upon n,m,k,p and δ, as above.
In the particular case where $k=m+1$ the proof is straightforward.∎

4.17. If $k>m+1$ then such an estimate for the Newtonian potential as in the foregoing Property 8 is impossible.

5. THE CASE WHERE Ω IS OPEN AND BOUNDED

5.1. Consider again the problem

$$D^k u = f \text{ in } \Omega \tag{1}$$

where now Ω is assumed to be open and bounded. Let $f \in L_p(\Omega;A)$ and define an extension $\tilde{f}$ of f to $\mathbb{R}^{m+1}$ by

$$\tilde{f}(x) = \begin{cases} f(x), & x \in \Omega \\ 0, & x \in \text{co}\Omega. \end{cases}$$

Then clearly $\tilde{f}\in L_p^{comp}(R^{m+1};A)$.

Now

$$E_k*\tilde{f} = \int_{R^{m+1}} E_k(x-y)\tilde{f}(y)dy^{m+1}$$

$$= \int_{\overline{\Omega}} E_k(x-y)\tilde{f}(y)dy^{m+1}$$

$$= \Phi_{\overline{\Omega}}$$

where $\Phi_{\overline{\Omega}}$ is the Newtonian potential with density $\tilde{f}$.

Accordingly to Property 3 the function

$$E_k*\tilde{f}$$

is a *weak solution* of

$$D^k u = f \text{ in } \Omega$$

which moreover enjoys the Properties 1,2,5 and 8.

5.2. As to the *strong* solutions of (1) the following two observations can be made.

First if it is possible to extend the given function f to a function f* which is $Lip_\alpha(\overline{\Omega};A)$ then it follows from Property 4 that

$$E_k*f^*_{\overline{\Omega}}$$

is a strong C_k-solution of

$$D^k u = f \text{ in } \Omega.$$

Secondly if the given right-hand side f is in $C_r(\Omega;A)$ with $r\geq 1$ it then follows from Corollary 3.7 that

$$E_k*\tilde{f}$$

is a strong C_{r+k-1} solution of (1).

5.3. If $f\in L_1^{loc}(\Omega;A)$ then, in general, this function cannot be extended to a distribution in R^{m+1}. This will be the case if and only if there exists a differential operator ∂^β and a function h which is continuous in $\overline{\Omega}$ such that

$$f = \partial^{\beta} h \text{ in } \Omega,$$

or equivalently if and only if f can be extended to a distribution T of slow growth in Ω : $T \in S^*(\Omega;A) = \mathcal{D}^*(\overline{\Omega};A)$. So assume that $f \in L_1^{loc}(\Omega;A)$ is extensible to a distribution $\tilde{T}_f \in \mathcal{D}^*(R^{m+1};A)$. Then we have in distributional sense

$$D^k(E_k * \tilde{T}_f) = \tilde{T}_f \text{ in } \mathcal{D}(R^{m+1};A)$$

and

$$D^k(E_k * \tilde{T}_f) = f \text{ in } \mathcal{D}(\Omega;A).$$

This means that

$$g = E_k * \tilde{T}_f$$

is a *distributional* solution of (1).
But then we know from Theorem 3.6 that g is in fact a *weak* solution which is in $L_{q_0}^{loc}(\Omega;A)$ with $1 \leq q_0 < \frac{m+1}{m+1-k}$ if $k < m+1$ and $q_0 = +\infty$ if $k \geq m+1$.

5.4. If $f \in L_1^{loc}(\Omega;A)$ and f is not of slow growth in Ω then a solution to (1) can be constructed in

$$\Omega_\varepsilon = \{x \in \Omega : d(x, co\Omega) > \varepsilon\}$$

for any $\varepsilon > 0$.
With the same mollifier α_ε as in 3.2 we have indeed

$$\alpha_\varepsilon f \in L_1^{comp}(R^{m+1};A).$$

So in distributional sense

$$D^k(E_k * \alpha_\varepsilon f) = \alpha_\varepsilon f$$

from which it follows that

$$\Phi_{\overline{\Omega}_\varepsilon} = E_k * \alpha_\varepsilon f$$

is a *weak* solution of (1) in Ω_ε for each $\varepsilon > 0$, which enjoys the properties 1,2,5 and 8 of the Newtonian potential.

Acknowledgement

We are indebted to Prof.Dr. R. Delanghe for his encouragement and valuable comments.

References

[1] R. Delanghe and F. Brackx, Hypercomplex function theory and Hilbert modules with reproducing kernel, Proc. London Math. Soc. (3), 37(1978), 545-576.

[2] R. Delanghe and F. Brackx, Regular solutions at infinity for a generalized Cauchy-Riemann operator, Simon Stevin, 53 (1979), 13-30.

[3] R. Delanghe and F. Brackx, Duality in hypercomplex function theory, J. Functional Analysis, 37 (1980), 164-181.

[4] R. Delanghe and F. Brackx, Runge's Theorem in Hypercomplex Function Theory, J. Appr. Theory, 29 (1980)

[5] H.G. Garnir, Problèmes aux limites pour les équations aux dérivées partielles de la physique (Université de l'Eat à Liège, Cours de Troisième Cycle, 1975-76).

[6] V. Iftimie, Opérateurs du type de Moisil-Teodorescu, Bull. Math. de la Soc. Sci. Math. de la R.S.R., 58 (1966), 271-305.

[7] F. Sommen, A product and an exponential function in hypercomplex function theory, to appear in Applicable Analysis.

[8] F. Sommen, Spherical monogenic functions and analytic functionals on the unit sphere, to appear.

[9] I.N. Vekua, Generalized analytic functions (Perganon Press, Oxford, 1962).

ABCDEFGHIJ–AMS–898765432